U0111349

陳家廚坊
Chan's Kitchen

陳家廚坊
Chan's Kitchen

SHANGHAI
CUISINE

品味上海菜

方曉嵐・陳紀臨 著

陳家廚坊
Chan's Kitchen

萬里機構

序

　　記得在五、六十年代的香港，凡是講話帶有江浙口音的人，都被叫做上海佬、上海婆，而當年香港有不少上海館子，一律叫做上海舖。很懷念那個年代的上海舖，店舖都不大，店前的明檔擺滿了一個個的冷盤熟食盤子，五香燻魚、油爆蝦、醉雞、鳳尾魚、鹽水毛豆、醬蹄、西芹豆乾、五香花生，林林總總任君選擇。不能叫做涼菜或冷盤，因為食物的溫度是暖的，當天做好放着賣，不冷也不熱，只好叫做冷盤熟食。

　　我母親是浙江紹興人，小時候，我和姐姐下課後，常常跟着母親上江浙館子吃午飯，反而比較少上茶樓吃點心，飲茶只是週末才有的節目。記得那時，母親帶着我們到學校附近的上海舖，幾乎必吃的是油豆腐粉絲湯和嫩雞煨麵，還有上海粗炒，外加一兩小碟冷盤熟食。派我們小孩子去選揀冷盤熟食，是母親給予的恩惠，我和姐姐都會感到很興奮，跳到櫃前商量半天，結果也只是准拿兩款，媽媽的眼睛在我們腦袋後面盯着呢！

　　很喜歡上海，舊傳統與新潮流融合在一起，令人沉醉，受母親的影響，我自小便愛吃上海菜。九十年代，我們因工作需要，常常要到蘇州和常熟，來往旅程故意停留上海，是心領神會的理由。飛機上，就算餓透了，也不會吃那盤便餐，拿個麵包只為了防患未然，擔心會遇上公路大塞車，那時的上海，塞車情況嚴重，一個半小時能到酒店算是福星高照。在路途上，心裏一直在盤算着晚餐到哪裏能據案大嚼一頓豐富的上海菜，由早上的豆漿燒餅開始，每天起碼吃三至四頓，飛機上拿的那可憐的小麵包，總是必然地躺在酒店的垃圾桶中。

後來去過很多次上海，也曾小住過，看了樓盤，幾乎下訂簽約，但突然其來的冷靜，結果沒有買下來，錯！錯！錯！現在樓價升了好幾倍，原來有些事還是衝動的好。上海一直在變，一日千里的進步，變得認不出了，浦東機場、上海磁浮列車、高鐵、偌大的商場、高科技的大樓……數不盡的繁華先進，但不變的，是我對上海味道的懷念，還是那樣的魂牽夢縈。

過去十年，曾多次提筆想寫一本上海菜的食譜書，幾年前就寫好了出版大綱，但考慮到上海菜的一些特別的食材，也只在秋天才當造，例如大閘蟹、蠶豆、馬蘭頭、草頭和薺菜，還有11月才有的塌菜；每年為了新書能參加七月書展，我們通常的拍攝工作都安排在每年四月，但四月買不到這些秋天的食材，加上秋天是旅行的好季節，往往總是呼朋喚友遊玩而無心向學，因此也就一直與出版上海菜食譜書無緣。2020年，是特別的一年，新冠肺炎至今擾攘了大半年，還是反反覆覆不肯離開，本來在七月的書展，一再延期至明年七月舉行，這就是天意，正好讓我們安心在家撰寫和拍攝上海菜，了卻這多年的心願！

在這特別的日子，謹以此書，獻給所有為抗疫而辛勤工作的醫護人員，還有千千萬萬，乖乖地留在家中抗疫的你和我。

祝願大家身心健康，惜福！感恩！

方曉嵐

2020年12月

目錄

上海 · 尋味

　　上海是中國最大的城市、直轄市、國際金融中心和航運中心。上海位於中國東部長江三角洲，面向太平洋，優越的地理位置，造就了上海幾百年來成為中國南北海運的要衝，以及長江流域各省與世界交往的門戶。

　　上海菜，是中國菜系中重要的組成部分。上海屬亞熱帶濕潤季風氣候，雨量充沛，四季分明，土地肥沃。兩千多年前，這片土地是長江出海口的一個東海邊上的漁村，更是古代海鹽的重要產地。在這片臨海的魚米之鄉土地上，有着兩千多年的農耕和漁業的歷史，特殊的地理位置和良好的氣候條件，為上海菜的發展提供了優良的食材條件，沉澱着純樸而濃厚的鄉土風味。

上海是古代東周王朝四公子之一楚國春申君的領地，隋朝時設華亭鎮，宋代時設上海鎮，是「居海之上」的意思。元朝時設立為上海縣；明代開始，上海縣日漸興旺，並成為江南的絲綢及綿紡集散中心，百貨雲集，商貿發達；清初時，朝廷曾將上海縣併入江南省；到了乾隆時期，把江南省重分成江蘇、浙江、安徽，而上海縣當時屬於江蘇省；1843 年，在鴉片戰爭後，清政府實行五口通商，上海縣是五個通商口岸之一，上海正式開埠；民國時朝，上海縣脫離江蘇省，成為上海市，後成為中國的直轄市。

上海是一個既古老又富傳奇色彩的城市，上海人生活得精緻考究，獨特的大城市文化，素有「東方巴黎」之稱。上海多年來的經濟繁榮，孕育出別具一格的上海菜，稱為滬菜，特點是在菜式上海納百川，既不失傳統的鄉土風味，又集江南各大名菜之大成，成為馳名中外的中國菜流派之一。

上海菜的形成和興起，與上海的城市發展息息相關。17 世紀之前，上海菜只有本幫菜，源自上海開埠前，農村人在鄉鎮或進城途中擺賣的「飯攤」，也即是聚集的吃飯排檔；後來，這些飯攤陸續進城開小飯店，賣的都是些老上海家常菜，只做小菜，沒有宴客大菜，「炒幾道小菜」是上海人的口頭禪。

1842 年，中英簽訂了「江寧條約」，上海開放港口通商，很快便成為中國華東的紡織和絲綢的貿易中心，引來了江南各省以至廣州等各地的藝術文化人、商人和各行業的求職者。上海在此段時期，為了解決新來移民的居住需要，新建了很多房屋，而大街上伸延的巷子愈建愈多，這些巷子叫做「弄堂」，大多數以往的飯攤，都遷入了弄堂中經營，為居民提供經濟實惠的家常菜。紅油赤醬的本幫菜，在弄堂中得到廣泛的認同。

本幫菜的特點是多油、味濃、糖重、色澤濃艷；採用紅醬油做紅燒菜式，濃油赤醬是上海菜的代表，味道偏甜偏濃，口感軟爛，汁稠芡厚；香味濃郁的精煉蔥油，更為菜式增加誘人的魅力。本幫菜選料豐富，包括雞、鴨、豬、牛、菇、

筍、河鮮、海產、豆製品、麵筋、醃菜、蔬菜等，傳統做法多為紅燒、煸炒、燴、糟、燉，以及長時間煨煮等烹調手法，充滿小城鎮的氣息。

異地移居上海的商人、文人，和隨之而來的廚師、廚娘、太太和姨太太們，為上海帶來了揚州、杭州、蘇州、無錫、紹興、寧波、南京、安徽、廣東等地的飲食文化，既與本幫菜平等存在，但又依然保留着故鄉的風味，這與多年來上海居民多樣性的祖籍構成有關。於是，「海幫菜」應運而生，又稱為「海派菜」，亦即「海派江南風味菜」。

海幫菜的口味，較傳統本幫菜清淡、輕油、薄芡，比較多使用生煸、滑炒、涼拌，刀工精細，配色講究，特別是善烹水鄉河鮮和新鮮蔬菜，受杭州菜和江蘇菜的影響，以多種野菜及水生植物入饌。

1949 年新中國成立之初，百廢待興，上海市首先致力恢復大城市風貌，因應城市的急速發展，加快建成了不少大酒店，很多山東魯菜名廚應聘到上海工作；1950 年代的上海餐飲名店，包括燕雲樓和國際飯店的主廚都是魯菜大師，他們把魯菜精美而多樣化的烹飪技巧，帶到了上海，培養了大批的本地廚師，更為上海的餐飲業引入了很多山東菜的名菜，令上海菜得到進一步多元化發展。

上海菜的發展，得益於吸收各地的風味菜，集百家之長，不斷進步，這是上海菜的優勢。近三、四十年的改革開放，作為直轄市的上海，經濟發展一日千里，餐飲業更迅速發展。上海是全國第一個城市率先進行烹飪理論研究，並強調以理論指導實踐，力求在繼承傳統的基礎上，不斷發展創新。

今天的上海菜，是本幫菜與海幫菜的混合，不分彼此，共同進步，菜式講求清新秀美、溫文爾雅、風味多樣；在延續它的精緻和考究的同時，向低脂、低糖、低鈉方向改良，就像它富有生命力的前世今生，上海菜充滿了新時代的氣息，繼續前行！

上海菜的四種必用技法

怎 樣 煉 豬 油

Making of lard

豬油在上海菜中是常見的油料，有些菜一定要用豬油，能夠增加菜式的香味，使菜式更加潤滑亮麗。例如本書中介紹的紅燒河鰻、響油鱔糊、生煸草頭、油醬毛蟹、銀魚跑蛋、苔菜年糕等菜式。煉豬油所用肥肉的部位，要選用豬板油或豬背油。

肥豬肉	300 克
蒜頭	2 瓣（拍開）
薑片	30 克

1. 肥豬肉洗淨切小粒。在鑊裏放 125 毫升水，放入肥豬肉，燒開，轉小火慢慢熬至豬肉粒呈焦黃色成豬油渣，撈出，再放入蒜頭、薑片，炸至變焦，撈出。

2. 豬油涼卻後呈白色，可放雪櫃保存，隨時可用。

＊英文食譜見第 165 頁

怎 樣 熬  葱 油

Making of spring onion oil

　　煮上海菜離不開葱香撲鼻的葱油，除了用來炒菜及做葱油拌麵和葱開煨麵外，很多上海菜在起鍋裝盤前都淋上葱油，以增加菜的香味。

 材料

葱	300 克
油	500 毫升

 做法

1. 葱洗淨，切成三段。

2. 先把油燒至中溫，放入葱白，用最慢火熬至葱白變軟變黃，撈出，再把其他的葱段放入，熬至乾身並呈焦黃色，撈出。

3. 葱油放涼後放冰箱保存。炸過的葱可用於葱油拌麵或葱開煨麵等菜式。

怎 樣 炒 糖 色

Making of caramel

　　炒糖色在做濃油赤醬的上海菜中，是不可缺少的一道工序，能夠為食物添加美麗的紅色，增進食慾。紅燒肉先用糖色炒製，後落醬油來燜，瘦肉部分更鬆軟。

材料

白糖	2 湯匙
水	3 湯匙

 做法

1. 在鍋中放入白糖和 1 湯匙水，用中小火煮溶白糖，邊煮邊攪拌，糖溶化後先是呈大泡沫，再慢慢變為小泡沫，這時要轉為小火，煮至糖漿的顏色變為深琥珀色，熄火。

2. 等 1 分鐘讓糖漿稍為涼卻，再把餘下 2 湯匙水徐徐的加入鍋中拌勻即成。

*注意：剛剛炒好的糖漿溫度非常高，立刻加水會濺起，所以要稍放 1 分鐘才加水。這是做一道菜用的糖色份量。也可以多做，放雪櫃保存。

怎 樣 製 香 糟 汁

Making of fragrant distiller's grain sauce

> 糟是江浙菜的一個特色,有了香糟汁,就可以很方便的在家裏做出糟雞、糟鴨、糟肉、糟豬肚、糟蝦、糟魚等人們喜愛的糟貨。

香糟汁

香糟泥

材料

香糟	150 克
薑	30 克(切片)
葱	3 條(切段)
鹽、糖	適量

做法

1. 把香糟、薑、葱和 1500 毫升水煮沸,轉小火煮 30 分鐘,熄火,泡 8 小時讓香糟沉澱。

2. 再用棉布過濾兩至三次,得約 750 毫升香糟汁,加 2 湯匙鹽和 3 湯匙糖調味,不用時放雪櫃保存。

3. 香糟汁可以循環再用一、兩次,但是每一次要加適量的鹽和糖,煮沸後放涼才能保存。鹽和糖的比例約為 2：3。

＊熬葱油、炒糖色英文食譜見第 165 頁

製香糟汁英文食譜見第 166 頁

11

品味上海菜　上海菜的四種必用技法

認識酒糟

酒糟

　　酒糟，就是釀酒時糧食發酵取得酒液之後，剩下來的渣滓經調製過濾之後，用作調味料，稱為酒糟。中國各地自古都有釀酒坊，各地用來釀酒的糧食都不同，產生了各有地方特色的酒，同時也產生了不同的糟。著名的酒糟種類大致上有三種，最常見的是杭州和紹興地區，用糯米釀製黃酒的酒渣做的黃糟，香味濃郁，含酒精大約 8%，跟煮菜用的普通紹興酒差不多。第二種是福建紅糟，瓶裝的紅糟又稱為紅糟醬，特色是在釀酒時，已加入 5% 的紅米，釀酒後的糟是紅色的，可增加菜餚的色、香、味。紅糟醬在福建、台灣和新加坡都很流行，廣泛用於菜餚之中，著名的菜式有紅糟肉、糟煎帶魚、香糟雞片等。第三種是山東黃糟，當地的酒是用黍米釀成，在黍米酒糟裏再加入 15% 的炒熟麥麩和五香粉，味道很獨特。

　　吊糟，是一種傳統的懷舊調味料，江浙人對糟汁的要求很高，古法製糟汁，要講究「吊糟」。「吊糟」在江南地區已有好幾百年的歷史，方法是將酒渣兌入上好的黃酒，再次發酵，產生一種更深層次的液體，稱為糟汁。所謂「吊糟」，矜貴在於用的是陳年酒糟和那人手操作的慢吊。有了陳年酒糟，便進入吊糟的時候。先把上好的花雕酒按份量配好，混合在酒糟中，加入冰糖、鹽、桂花滷混合，再一起倒入一個特別製造的大棉布袋，這個袋的棉布針數要求非常緊密，要把酒滓都隔在袋內，而汁液要很慢很慢的滲出，而且拉力要很強，重重的袋子被掛在一個大桶上面，酒糟和花雕緩慢地在袋中再融合、發酵，從袋子滲出來的是透明的、像琥珀色的液體，散發出一種特殊的幽香。經過七日七夜，袋子裏的糟汁才算流完。完全靠傳統工藝和耐性。吊糟的名貴和醇厚神韻，也就只有堅持高要求的廚師才能承傳下來。

香糟

包裝的香糟，是用來做上海的糟貨的材料，代替了工藝複雜的吊糟。香糟是由小麥加工製成，把小麥磨碎，加工發酵成麥麯，再釀製而成。香糟含酒精度約 20%，是中國江南地區一種特殊的調料。包裝袋內的香糟，是渣狀的穀物，呈泥土顏色，香味濃郁。把香糟加水和薑蔥煮成濃濃的汁，再經過幾次過濾和沉澱出來的香糟汁，用鹽和糖調味後是做糟魚、糟雞等糟貨的材料。包裝香糟在南貨店有售（見圖）。

糟滷

超市和南貨店有售的瓶裝糟滷，是香糟的再加工產品，使用方便，也比較衛生，在家做菜也可用糟滷調味代替香糟做糟貨。糟滷主要用於烹調菜餚，例如糟溜魚片、糟蒸鴨肝、糟煨肥腸等，也會用來調配醉雞、醉蟹、醉豬手的醉汁。

冷盤
熟食

很懷念五、六十年代香港的上海舖，店舖不大，店前的明檔擺滿了一個個的熟食盤子，燻魚、油爆蝦、醉雞、鳳尾魚、鹽水毛豆、醬蹄、西芹豆乾、五香花生，林林總總任君選擇。食物的溫度還是暖的，當天做好放着賣，不涼也不熱，只好叫做冷盤熟食。

四喜烤麩

SAUTÉED WHEAT
KAOFU
WITH VEGETABLES

很多人都以為烤麩是豆製品，但其實烤麩是小麥的製品。

由小麥磨粉做成麵筋，再發展出來的食材有很多，其做法、用法、風味、形狀都各有特色，大家所熟悉的，就是用來烹調齋菜的各種形狀的麵筋。烤麩的「麩」就是麥子皮的意思，傳統的烤麩就是用帶麥皮的麥子，磨成的高筋麵粉做成的麵筋，經過保溫發酵後，用大火隔水蒸熟再定型而成的一種食材。

烤麩在中國江南一帶的江蘇、浙江、安徽等省份都是很普遍的食材。烤麩其貌不揚，像一塊發黴的大鬆糕，在南貨店購買時才用刀切出部分來。香港南貨店賣的烤麩有兩種，一種叫做「上海烤麩」，像有氣孔的海綿大鬆糕，要用刀切開，口感較腍。另一種的紋理清晰，氣孔不多，可以用手撕成小塊而不需用刀，口感帶有韌性，這是比較傳統的舊式烤麩，可惜在國內已經很少買得到，幸好在香港還買得到，所以叫做「香港烤麩」。

樸實無華的四喜烤麩，是很多上海人的兒時回憶，也是最普遍的家庭常備菜。它的做法簡單，食材健康，可作前菜暖食，也可以熱吃。做四喜烤麩，有兩派不同的喜好，一派是要求烤麩煮腍入味，所以在炒烤麩時加少許水或湯來煮，上海人做的四喜烤麩，會用這一種燴煮的技法。第二派是要求烤麩保留一點炸過的脆口，所以不會加水煮，是用炒的技法。

烤麩有發酵的酸味，必須預先汆水，沖水後再擠乾，並重複兩次，確保把酸味洗去。

4 人份

準備時間
20 分鐘

烹調時間
10 分鐘

上海烤麩

香港烤麩

可以用手將烤麩撕成小塊

材料

烤麩	200 克
毛豆仁	少許
乾雲耳	5 克
小冬菇	6 朵
金針（黃花菜）	20 克
筍	50 克
薑汁	2 茶匙
薑絲	10 克
蠔油	1 湯匙
紹興酒	1 湯匙
糖	1 茶匙
鹽	1/2 茶匙
麻油	1 茶匙

做法

1. 烤麩撕成小塊，用水放 1 茶匙薑汁煮沸，放入烤麩汆水 1 分鐘撈起，
   沖水後用手擠乾水分，然後再重複這個流程一次。
2. 燒熱一鍋水，把毛豆仁沸煮 5、6 分鐘，瀝乾。
3. 用清水發好雲耳，在沸水中煮 2 分鐘撈起瀝乾。
4. 冬菇浸軟，去蒂。
5. 金針浸 10 分鐘，撈出沖水。
6. 筍切片，放沸水中煮 2 分鐘撈起瀝乾。
7. 用 250 毫升炸油燒至六成熱（約 160℃），放下烤麩炸成金黃色，撈
   起瀝油，再用鑊鏟把油分盡量壓出。
8. 留 3 湯匙油大火起鑊，爆香薑絲和冬菇，加入烤麩、毛豆仁、筍片、
   金針、雲耳同炒，鑊邊瀢下紹興酒，加入蠔油、鹽和糖炒勻，最後加
   麻油拌勻，即成。

＊英文食譜見第 166 頁

馬蘭頭拌香乾

KALIMERIS
AND
TOFU SALAD

馬蘭頭

馬蘭，菊科，又別名紅梗菜、雞兒腸、田邊菊、紫菊、魚鰍串、螃蜞頭草等，為多年生草本植物，葉可食用；原本是一種田間野菜，現在大多數都是種植的，真正野生的馬蘭頭，只有在上海才買到，價格比較貴。

馬蘭頭拌香乾，是一道上海春天時很受歡迎的菜式。上海人稱馬蘭的葉為「馬蘭頭」（見圖），是指馬蘭葉的頭段嫩葉可食部分的意思。清代袁枚的《隨園食單》中，說到馬蘭頭是這樣的：「馬蘭頭菜摘取嫩者，醋合筍拌食。油膩後食之，可以醒脾。」馬蘭頭纖維豐富，帶清香味，用水略燙過後，吃之味道甘而後味特強。馬蘭頭雖然味帶草香和苦味，但它是一種愈吃愈有韻味的蔬菜，進食時應要慢慢咀嚼，才能真正品嚐出馬蘭頭特有的滋味。

材料選購：馬蘭頭在香港的上海南貨店有售，最好是現買即用，不要留過夜，會影響香味。豆腐乾可買五香豆腐乾或普通的白豆腐乾。兩頭通的不銹鋼圓模（直徑 7 厘米，高 5.5 厘米）（見第 23 頁圖），可以在烹調用模具店買到。

4-6 人份

 準備時間
5 分鐘

 烹調時間
10 分鐘

材料

馬蘭頭	300 克
豆腐乾（香乾）	2 塊
麻油	1 茶匙
鹽	1/2 茶匙
杞子（裝飾用）	數粒

 烹調心得

1. 馬蘭頭在沸水中不要燙太久，否則會變色。

2. 豆腐乾本身是熟的，稍為在沸水中拖一下便可。

3. 沸水焯馬蘭頭時加點生油，是為了保持翠綠顏色。

4. 焯水後沖涼開水，可增加脆嫩，而且可以除去一些草腥味。

5. 7×5.5 厘米的圓形不銹鋼模剛好能夠容納已經擠乾水的 300 克馬蘭頭和兩塊豆腐乾。

兩頭通的不銹鋼圓模

做法

1. 馬蘭頭洗淨，煮沸水，加 1/2 茶匙油，把馬蘭頭稍燙一下取出，用冷開水沖一下，用手擠乾水分備用。

2. 豆腐乾洗乾淨，用沸水稍燙，撈出備用。

3. 馬蘭頭和豆腐乾用廚紙吸乾後，剁碎混合，再加入鹽和麻油拌勻。

4. 把拌好的馬蘭頭和豆腐乾放在一個兩頭通的不銹鋼模內壓實，吃前倒扣在碟子中，加上杞子裝飾即成。

＊英文食譜見第 167 頁

老雪菜

雪菜

雪菜豆瓣酥

BROAD BEAN PASTE
WITH PRESERVED
POTHERB MUSTARD

豆瓣酥，也就是蠶豆酥，上海人這個「酥」，並不是像香港的叉燒酥，或者台灣的鳳梨酥，這個「酥」字是形容它的香軟鬆化。

蠶豆，是上海人經常用的食材，新鮮的帶莢蠶豆在春季上市，但冷藏的蠶豆肉在香港的南貨店四季都有售。上海人叫蠶豆肉做豆瓣，豆瓣醬就是用蠶豆作為原材料做的，但不會叫做蠶豆醬。

不少人小時候都吃過五香蠶豆，帶着豆莢，用牙「卡擦」地咬開，裏面是香脆的蠶豆肉，口感有點硬，我不喜歡吃。前幾天，三陽泰張先生送來一大包青綠色的冷藏蠶豆肉，我便做了個久違了的老上海菜豆瓣酥，順手放上網與朋友分享。豆瓣酥的圖片引來了很多朋友的興趣，看到留言，才知道原來香港有很多人都不知道有這個上海菜，甚至有些人還以為豆瓣酥是個餅，也有不少人曾經吃過這道菜，知道很美味，但惋惜已久未嚐此菜，只怕有日會在香港消失了。

傳統的老上海家常菜，只有小菜，沒有宴客大菜，「炒幾道小菜」是上海人的口頭禪，舊時上海大戶人家的姨太太，為留住先生的心出盡八寶，為免吃飯時先生久等，做兩三道拿手冷盤熟食，先討好先生的胃。大戶人家的豆瓣酥，是一道冷盤，也可以暖食，但不會熱食，平民百姓的飯桌上就不大講究了。

註：患蠶豆症（又稱「G6PD 缺乏症」）人士絕對不適宜進食此菜。

4-6 人份

準備時間
10 分鐘

烹調時間
20 分鐘

材料

冷藏蠶豆肉	500 克
老雪菜梗	100 克
葱油（做法見第 10 頁）	3 湯匙
清雞湯	250 毫升
糖	2 茶匙
麻油	

烹調心得

1. 雪菜即雪裏蕻，菜市場上買的雪菜有兩種，一種是青綠色的雪菜，即鹽水浸漬的雪菜；另一種是顏色偏黃的老雪菜，即發酵過的雪菜，老雪菜顏色不好看，但味道香濃。

2. 老雪菜的菜葉，味道很鹹，做豆瓣酥不能用雪菜葉，也不會好看，只用梗切粒味道已足夠，更不用加鹽。

做法

1. 把蠶豆肉汆水，過冷河，瀝乾水分。

2. 老雪菜梗沖水後擰乾，再切成細粒。

3. 燒熱葱油，翻炒蠶豆，加 250 毫升雞湯，煮到蠶豆變軟時，先盛起約 1/5，餘下的蠶豆繼續翻炒，邊炒邊用叉子壓碎。

4. 加入雪菜梗粒和已盛起的蠶豆，加糖，炒至收水成蓉，放至涼。

5. 用 1 個碗或食物盒，內壁抹上麻油，把蠶豆蓉放入，壓平表面，放冰箱 1 晚，吃時倒置在碟中，即成。

＊英文食譜見第 167 頁

燻魚

| SPICED SMOKE FISH

夜半酣酒江月下，美人纖手炙魚頭。

——清·鄭板橋

　　中國烹飪法中的「燻」，是將醃過的生料或熟料，用米、糖、蔗渣、茶葉、樟葉、松葉、果木等燃料，通過燻爐產生煙，把生或熟的原料燻至熟，或燻至着色及入味。上海的燻魚做法，源自江蘇，又稱為蘇式燻魚，製作方法是油炸浸滷，是中菜中少有的沒有經過煙燻過程，但卻被稱為「燻」的菜式。

　　在很久很久以前，燻魚的做法比較精緻講究，把炸酥了的魚塊放入醬汁中吸味之後，再置於鐵絲網上，用白炭小火慢烤，並不斷刷上醬汁，使其顏色一致，醬汁稍乾成焦糖，這才成為真正的燻魚，不過這種做法現在已經很少人會做了。

　　一般用肉質厚實的魚類製作燻魚，包括鯇魚、草魚、青魚等淡水魚，比較講究的宴客菜，就用鯧魚來做燻魚，用日本鯖魚來做效果也甚佳。選擇皮下脂肪比例較厚的魚類，味道會特別甘甜。在上海的餐館及街上的熟食店，都有出售真空包裝的燻魚，是遊客們喜愛的手信之一。

4 人份

準備時間 15 分鐘

烹調時間 15 分鐘

材料

鯇魚前段（不帶魚頭）	500 克
葱（葱白切段）	2 根
薑蓉	30 克
鹽	1/2 茶匙
紹興酒	2 湯匙
上海特級醬油	2 湯匙
五香粉	1/2 茶匙
鎮江醋	2 湯匙
片糖	1.5 塊（壓碎）
麻油	2 湯匙

上海特級醬油

做法

1. 鯇魚前段橫切成厚片，用乾淨布或廚紙吸乾水分。
2. 把葱白拍扁，放入大碗內，加入薑蓉、鹽和 1 湯匙紹興酒，放入魚塊拌勻，要確保每一塊魚都沾上醃料，醃製 15 分鐘。
3. 在另外的小鍋加入 1 湯匙紹興酒、醬油、五香粉、鎮江醋、片糖和 4 湯匙水，再加入醃魚的汁和葱青的部分，用小火煮至片糖完全溶化，用大碗盛起。
4. 魚塊醃好後，用廚紙吸乾醬汁，用 1000 毫升炸油中火炸至金黃色，瀝去多餘的油分。
5. 趁熱把剛炸好的魚塊逐一放入蘸汁中稍浸一下，吸收汁味，然後拿出放在碟上，再掃上麻油即成。

* 英文食譜做法見第 168 頁

上海的糟貨

很多人喜歡吃醉雞，但上海菜中的糟雞，是年輕一代廣東人和香港人比較少知道的美食。我們曾經做了一碟醉雞和一碟糟雞，放在一起給朋友們試吃，結果是所有人都大讚糟雞好吃，有朋友甚至說，若有糟雞，就不再吃醉雞了。「醉」是聞得到吃得出的酒味，擺明車馬；「糟」是吃不到酒味，卻有深藏不露的酒香，低調而有韻味。

鄰近上海的浙江紹興市，是黃酒之鄉，也是糟貨的發源地。糟貨的意思，就是各種用香糟浸製的食物。自從 1930 年代開始，上海人在夏天興起吃糟貨喝小酒的風氣。當年上海南京路上的杜五房、杜六房，還有滬東、滬西的兩間狀元樓，還有淮海中路的老人和、西藏中路的馬永齋等，都是著名的糟貨店，如今時移世易，也不知道這些名店還存在否，但肯定上海現在還有不少的糟貨店。

上海的糟貨店，品種百花齊放，有葷有素，包括糟雞、糟鴨、糟鵝、糟蟶子、糟黃魚、糟帶魚、糟茭白，糟豬舌、糟豬肚、糟豬尾。糟豬爪（豬手）、糟鳳爪、糟素雞等等，據說有上百品種，去到糟貨店，簡直令人目不暇給，垂涎三尺，樣樣都想買，每樣買一點就湊成一大盤。現在上海人夏天吃糟貨喝小酒的習慣還是流行的，特別是電視轉播奧運和足球賽的日子，一盤香味濃郁的什錦糟貨，加上幾支凍啤酒，說不出有多愜意。

THREE TIDBITS
IN FRAGRANT DISTILLED
GRAIN SAUCE

糟三樣

糟　糟　糟
蝦　豬　毛
　　肚　豆

糟毛豆

FRESH SOY BEANS IN
FRAGRANT DISTILLED
GRAIN SAUCE

材料

毛豆	300 克
花椒粒	1 湯匙
鹽	1 茶匙
五香粉	1 茶匙
香糟汁（做法見第11頁）	

做法

1. 帶莢新鮮毛豆沖洗乾淨瀝乾，用剪刀剪去毛豆莢的頭尾兩端。

2. 大火煮沸半鍋清水，放入花椒粒、鹽、五香粉，煮5分鐘後，放入剪好的毛豆，水的份量要稍為浸過毛豆，用中火煮20分鐘熄火，再讓毛豆浸在鍋中10分鐘，撈出毛豆放涼，煮豆的水和香料都不要。

3. 用廚紙吸乾毛豆莢上的水分，再放入香糟汁浸着，放在冰箱中約4至5小時，中途兩次用筷子把毛豆翻一下，使每粒毛豆都浸到香糟汁。

＊英文食譜見第 168 頁

材料

豬肚	1 個
生粉	2 湯匙
鹽	1 湯匙
白醋	2 湯匙
白胡椒粒	2 湯匙（拍碎）
香糟汁（做法見第 11 頁）	

做法

1. 把新鮮豬肚從內翻出，用生粉抓洗，沖淨，再用鹽抓洗，沖淨，剪去附在豬肚壁內的油，再翻過來，洗淨，切成幾大塊。

2. 燒熱一鍋水，放入豬肚，加白醋和白胡椒，大火沸煮 15 分鐘，轉中小火煮 1.5 小時至腍，撈出，用水沖洗乾淨，瀝乾，斜切成條。

3. 把豬肚放進香糟汁，浸泡 24 小時即成。

＊英文食譜見第 169 頁

糟豬肚

PIG'S STOMACH IN
FRAGRANT DISTILLED
GRAIN SAUCE

糟蝦

PRAWNS IN FRAGRANT
DISTILLED GRAIN SAUCE

基圍蝦⸺⸺⸺⸺⸺⸺⸺⸺ 300 克
香糟汁（做法見第 11 頁）

做法

1. 基圍蝦沖洗乾淨，用水灼熟後，瀝乾
   水，放涼。
2. 剪去蝦鬚、蝦腳，泡在香糟汁內 4 至
   5 小時後可食。

＊英文食譜見第 169 頁

酒糟鴨舌

DUCK TONGUE
WITH
PICKLED SAUCE

　　中國江南地區河道湖泊甚多，有很多養鴨人家，更建有不少現代化的大型宰鴨工廠，生產很多不同的鴨產品，有出產南京板鴨和鹽水鴨等整隻光鴨，也有分拆了的冷藏鴨腿、鴨翅膀（鴨翼）、鴨腳、鴨脖子、鴨內臟、鴨胸肉和鴨舌，使我們烹調材料的選擇比以前豐富了很多。

　　廣東人叫鴨舌做鴨脷，它是一種奇妙的東西，吃到口裏是軟軟的，稍為有點彈性，但不像其他豬舌和牛舌有那種非常結實的口感，在眾多禽類中，好像只有鴨子的舌頭被用作入饌之用。鴨舌易熟，不用煮太久，以保持脆爽的口感。

　　鴨舌本身沒有任何味道，亦談不上有甚麼營養價值，但是它的柔軟組織，卻是各種味道最好的載體，可塑性很高，用來炒、滷、醬、醉、糟、炸都有很好的效果。

　　用糟滷來做鴨舌，做法比較簡單。糟滷（見第 14 頁）在上海南貨店或大超市有售。喜歡味道濃的話，可把鴨舌浸的時間延長到半天，但糟滷味道較鹹，如果時間過長，鴨舌可能會過鹹。

品味上海菜 ❀ 冷盤熟食

🍲 4 人份	浸泡時間 3 小時
準備時間 10 分鐘	烹調時間 45 分鐘

材料

鴨舌	300 克
花椒	30 粒
八角	2 粒
薑	4 片
糖	1 茶匙
鹽	1/2 茶匙
蔥	3 根（切段）
紹興酒	15 毫升
糟滷	50 毫升
麻油	1/2 茶匙

做法

1. 把鴨舌洗乾淨，撕去舌衣，在沸水中煮20分鐘，再用冷水沖洗，瀝乾水分。
2. 把花椒八角放在小紗布香料袋中。
3. 用 500 毫升水，加入薑片、香料袋，煮 20 分鐘，薑片和香料袋取出不要。
4. 煮好的滷水加入糖、鹽、蔥段，煮沸後熄火，涼卻後加入紹興酒和糟滷，製成糟汁。
5. 把煮熟的鴨舌用糟汁浸泡約 3 小時後可撈出，再淋上麻油即可。

＊英文食譜見第 170 頁

品味上海菜 ❀ 冷盤熟食

糟雞

CHICKEN MARINATED
IN FRAGRANT DISTILLED
GRAIN SAUCE

客醉眠未起，主人
呼解醒。已言雞黍熟，
復道甕頭清。
——唐・孟浩然
《戲贈主人》

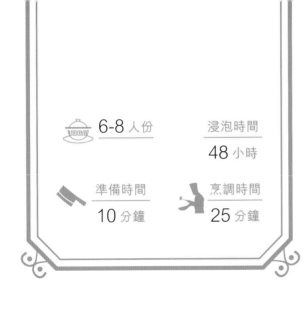

6-8 人份 | 浸泡時間 48 小時

準備時間 10 分鐘 | 烹調時間 25 分鐘

做法 ·························

1. 把雞洗淨後瀝乾。

2. 燒沸一鍋水,放下雞,雞胸朝上,再煮沸後加蓋,熄火,把雞浸約 17 分鐘;把雞翻轉,雞胸向下,再浸 5 分鐘,取出放涼,切成兩邊。

3. 把雞和香糟汁放在一個大食物密實袋裏,擠出空氣,放冰箱裏 24 小時。

4. 把雞從袋中取出,斬件後淋上少許香糟汁即可食用。

＊英文食譜見第 170 頁

品味上海菜 ✿ 冷盤熟食

材料

新鮮雞 ·········· 1 隻(約 1200 克)
香糟汁 ·········· 300 毫升
　　(做法見第 11 頁)

蜜汁鳳尾魚

DEEP FRIED GRENADIER ANCHOVY

鳳尾魚，又名鳳鱭，屬鯷魚科，是刀魚的親戚，但身價就差天共地了。七、八十年代，香港流行吃江浙菜，江浙菜館叫做「上海舖」。當時街上的大小「上海舖」，都有一個非常精彩的冷盤櫃，展示着各式江浙冷盤熟食，有油爆蝦、炸鳳尾魚、雪裏蕻毛豆、油燜筍、糖醋小排骨等等，客人可以隨意挑選。看得眼花撩亂，吃得心滿意足。時移世易，現在滿街都是韓國菜日本菜，卻還是很懷念兒時那精彩的「上海舖」，還有那炸得香脆的蜜汁鳳尾魚。

4 人份

準備時間　25 分鐘
烹調時間　5 分鐘

材料

鳳尾魚	600 克
蒜頭	2 瓣
薑	10 克
鹽	1/2 茶匙
胡椒粉	1/4 茶匙
糖	3 湯匙
魚露	1 茶匙
紹興酒	1 湯匙

 烹調心得

1. 鳳尾魚細小，炸時要用大火，在最短時間內炸脆。火如果不夠猛，鳳尾魚就不會炸脆。
2. 每年春天，香港的魚檔就有鳳尾魚出售，到了四月底就全部失蹤，當造的時間不長，若有緣見到就不要錯過。

 做法

1. 蒜頭去衣剁蓉，薑去皮磨蓉，備用。
2. 鳳尾魚切去魚頭和魚腹，洗淨。
3. 用鹽和胡椒粉拌勻鳳尾魚，醃 20 分鐘，用廚紙吸去水分，抹乾。
4. 大火燒熱炸油 500 毫升，把鳳尾魚炸至脆身，取出備用。
5. 燒熱 1 湯匙油，放薑蓉、蒜蓉爆香，加糖、2 湯匙水和魚露，煮至收汁，放下鳳尾魚，灒紹興酒，拌勻即成。

＊英文食譜見第 171 頁

品味上海菜 ✿ 冷盤熟食

鹽水鴨

SALT WATER DUCK

白鵝炙美加椒後，錦雉羹香下豉初。箭茁脆甘欺雪菌，蕨芽珍嫩壓春蔬。

——南宋・陸游《飯罷戲示鄰曲》

　　上海菜是肚大能容、廣納百川。鹽水鴨是同處華東的南京市名菜，當然也被上海菜所吸納，收為己用。現在上海的食品店中，也有包裝的鹽水鴨出售。

　　南京市是中國六大古都之一，蘊藏着豐富的文化遺產，古意盎然。戰國時代，南京和蘇州同屬吳國，兩地的菜式風味雖有類同，但卻是各自發展。歷史上南京曾歸入楚國，受楚國（湖北）的影響，所以南京菜沒有蘇州菜那麼甜，而更着重的是鹹淡適中、原汁原味、醇厚而不膩。

　　南京人的飲食，總是離不開鴨子，城中大街小巷都有賣鴨子的食店，由小販到超市，由街頭小店到大酒家都在賣鴨子，南京板鴨、鹽水鴨、醬鴨、燒鴨，以及各種鴨血、鴨內臟，款式之多，真是令人目不暇給；這要歸功於江蘇省水道縱橫，河塘滿佈，自古就是養鴨子的好地方，特別是南京郊區的江寧縣，那裏生產的鴨子肉質肥美嫩滑，是南京人吃鴨子的首選。

　　上海的鹽水鴨，與南京的鹽水鴨製法相似，先醃後煮，鴨身油潤光亮，味道甘腴鹹香，肉質細嫩，是下酒的上佳菜式。

8 人份	醃製時間 48 小時
準備時間 15 分鐘	烹調時間 90 分鐘

材料

冰鮮嫩米鴨	1 隻
（淨重 1500 克）	
鹽	3 湯匙
花椒碎	1 茶匙

滷水材料

八角	2 粒
甘草粉	1 茶匙
草果	2 粒
花椒碎	1 湯匙
陳皮	1 角
丁香	6 粒
香葉	6 片
白胡椒粒	1 茶匙
薑	20 克（切片）
鹽	150 克
紹興酒	3 湯匙

做法

1. 切掉鴨尾巴，洗淨，吊起瀝乾水分。
2. 炒熱 3 湯匙鹽，加入花椒碎炒香，把鴨身內外抹勻，醃 6 小時。
3. 把滷水材料裏的香料放入香料袋中捆紮好。
4. 在大煲中放入香料袋、薑和 2500 毫升清水，煮沸後改小火煮 20 分鐘。
5. 加入 150 克鹽和 3 湯匙紹興酒，大火煮至鹽完全溶解。
6. 放入米鴨，鴨胸朝下，煮沸，再用小火煮 30 分鐘。
7. 把鴨翻過來，鴨胸朝上，再煮 30 分鐘。
8. 把鴨取出，放涼後斬件排在碟上，即成。

＊英文食譜見第 172 頁

 烹調心得

1. 鴨尾巴上的鴨羶子必須切去，否則味道會腥臊。
2. 鴨的烹調時間要按鴨子的大小和老嫩程度而作調節。

家常菜

雪沫乳花浮午盞，蓼茸蒿筍試春盤。人間有味是清歡。

——宋·蘇軾《浣溪沙》

濃油赤醬的弄堂小菜，是上海菜的根基所在，味道偏甜偏濃，口感軟爛，汁稠芡厚；香味濃郁的精煉蔥油，更為菜式增加誘人的魅力。

這章節的家常菜式選料豐富，包括雞、豬、菇、筍、河鮮、海產、豆製品、麵筋、醃菜、蔬菜等。

燒兩道美味小菜，道盡了上海人對在家吃飯的濃情厚意。

紅燒獅子頭

BRAISED MEATBALL,
HANGZHOU STYLE

紅燒獅子頭，是一道很受歡迎的上海菜餚。相傳一千多年前，隋煬帝楊廣三下揚州，帶着妃嬪隨從，乘着龍舟沿大運河南下看瓊花，「所過州縣，五百里內皆令獻食」。楊廣看了瓊花，特別對萬松山、金錢墩、象牙林、葵花崗四大名景十分留戀。回到行宮後，楊廣吩咐御廚以上述四景為題，製作四道菜餚，作為紀念。御廚們費盡心思終於做成了松鼠桂魚、金錢蝦餅、象牙雞條和葵花斬肉這四道菜。楊廣品嚐後，十分高興，於是賜宴群臣，一時間淮揚菜餚傾倒朝野。

到了唐代，隨着經濟繁榮，官宦權貴們也更加講究飲食。有一次，郇國公韋陟宴客，府中的名廚韋巨元也做了揚州的這四道名菜，並伴以山珍海味、水陸奇珍，令座中賓客們嘆為觀止；當「葵花斬肉」這道菜端上來時，只見那巨大的肉團子做成的葵花心，精美絕倫，有如雄獅之頭。賓客們趁機勸酒道：「郇國公半生戎馬，戰功彪炳，應佩獅子帥印。」韋陟高興地舉酒杯一飲而盡，說：「為紀念今日盛會，『葵花斬肉』不如改名『獅子頭』。」奉承者一呼百諾，從此就添了「獅子頭」這道名菜。

揚州菜以刀工細膩出名，做得好的揚州獅子頭是要入口即溶，肥而不膩，同時要能保持材料的原汁原味。古代的獅子頭用的豬肉是瘦三分肥七分，後來慢慢變成瘦四肥六，到了清代，又變了瘦肥各半。現代人講究健

康，瘦肥的比例又要改了。其實，在獅子頭的製作過程中，大部分的肥油已經流失了，所以入口能夠肥而不膩。

做獅子頭的材料，除了有豬肉外，可以加入鮮蝦、蟹肉或蟹粉等，但是重要的是不要讓這些材料蓋過了豬肉本身的鮮味。刀工方面，講究的是「細切粗斬」，也就是說先把肉切得很細，然後再加粗剁，做成大肉丸，就能顯出獅子頭毛髮蓬鬆的形狀，這樣外觀和口感都兼顧了。

紅燒的定義，是把材料作二次加工，先用炸、煎的方法把材料煮到定型，及起碼有七、八成熟，然後以湯水及深色汁醬煮至適度酥軟及入味，收汁埋芡，使濃汁附在材料上，做成口感柔滑外表紅亮的菜式。

6 人份

準備時間
20 分鐘

烹調時間
15 分鐘

材料

獅子頭材料

絞瘦豬肉 ⋯⋯⋯⋯⋯ 300 克
肥豬肉 ⋯⋯⋯⋯⋯ 200 克
麥片 ⋯⋯⋯⋯⋯ 1 湯匙
薑汁 ⋯⋯⋯⋯⋯ 1 湯匙
鹽 ⋯⋯⋯⋯⋯ 1/2 茶匙
糖 ⋯⋯⋯⋯⋯ 1/2 茶匙
胡椒粉 ⋯⋯⋯⋯⋯ 1/4 茶匙
生粉 ⋯⋯⋯⋯⋯ 1 湯匙

紅燒材料

冬菇 ⋯⋯⋯⋯⋯ 6 朵
薑片 ⋯⋯⋯⋯⋯ 6 片
紹興酒 ⋯⋯⋯⋯⋯ 1 湯匙
紅燒醬油 ⋯⋯⋯⋯⋯ 1 湯匙
糖 ⋯⋯⋯⋯⋯ 1/2 茶匙
蠔油 ⋯⋯⋯⋯⋯ 1 湯匙
清雞湯（或水）⋯⋯⋯ 125 毫升
麻油 ⋯⋯⋯⋯⋯ 1 茶匙
生粉 ⋯⋯⋯ 1 茶匙（埋芡）

紅燒醬油

做法

1. 把絞瘦豬肉加工再剁，一半粗斬，一半剁細。
2. 肥豬肉切成條，然後切粒剁細。
3. 麥片用手捏成粉狀。冬菇泡軟去蒂，切厚片。
4. 把豬肉、薑汁、鹽、糖、胡椒粉、生粉、碎麥片放大碗裏，用手拌到完全混合，再用手拿起向碗裏重撻幾下增加彈力，放在冰箱內冷藏 2 小時。
5. 把肉取出平均分成六份，用手黏水，把肉做成六個大丸子，再沾上薄薄的一層生粉。
6. 把炸油燒至中溫，放入肉丸，炸至表面略硬，轉小火，再炸至七、八成熟。
7. 倒出炸油，只留 1 湯匙油，用中火爆香薑片和冬菇，潛酒，加入醬油、蠔油、糖、放入肉丸上色，再加入清雞湯和泡冬菇水，煮沸後再煮約 2 至 3 分鐘，拿出肉丸，放在碟子中。，
8. 旺火收汁，用生粉埋芡，加麻油兜勻，淋在肉丸上，旁邊伴以青菜。

＊英文食譜見第 173 頁

品味上海菜 ❀ 家常菜

 烹調心得

1. 以上份量是六人份，每人吃一個，如果人數增加一兩人，也可以做成八個較小的肉丸，但如果做大了丸子，紅燒的時間就要增加，以免肉丸裏面不熟。
2. 加入麥片是要增加黏度，使肉丸不易散開，同時也可以吸收部分的肉汁。

醃篤鮮
| YANDUXIAN SOUP

　　江南的冬天又濕又冷，一家人圍着吃飯，最窩心的就是來一個熱呼呼的砂鍋醃篤鮮。這個菜式據說起源於清代，醃篤鮮在杭州又叫做「南肉春筍」，因為杭州人把鹹肉叫做「南肉」。醃篤鮮是中國江南傳統的開春名菜，流行吃醃篤鮮的地區，包括上海、江蘇、浙江，難以準確地說是最早源自哪個地方。

　　醃篤鮮有三個主要材料：上海鹹肉、春筍、豬肉。清代著名文學家美食家李漁曰：「肉之肥者能甘，甘味入筍，則不見其甘，但覺味至鮮。」意思是認為煮筍必須配豬肉，而且要帶肥的豬肉，味道才會更鮮美。

　　竹筍，就是竹的嫩芽，每年農曆二、三月，正是春筍當造的季節，在杭州天目山竹林蔽天，出產的竹筍，嫩滑無渣爽脆鮮甜，是春筍中之極品。杭州的家鄉南肉（鹹肉）跟金華火腿一樣，都是用金華兩頭烏的豬肉醃製而成，但肉的部分不同，金華火腿用的是豬後腿，鹹肉是豬五花肉或肋條肉，金華火腿要醃製三年，而鹹肉醃製時間就短得多。

　　醃篤鮮這個名字其實一點也不古怪，而且對菜式的表達十分直接清楚，「醃」是代表鹹肉，「篤」是指這個湯菜是用長時間慢火燉，湯面在沸煮時發出「篤、篤、篤」的聲音，就像廣東方言說的：「滾到卜萄聲」，但也有人說「篤」字是指冬筍，因為兩個字都同為竹字頭。醃篤鮮的「鮮」就是鮮豬肉，再加上春筍、小棠菜這些時令新鮮蔬菜，就成了口味鹹鮮、湯白味濃、筍香鮮嫩、鹹鮮兩肉酥爛甘香的醃篤鮮。

6 人份

準備時間 20 分鐘　　烹調時間 2.5 小時

 做法 ....

1. 把鹹肉、五花肉汆水 3 分鐘，取出洗淨。
2. 薑切片，葱打結。小棠菜修剪留嫩心部分，用滾水略燙，立即拿出用冷水沖洗，以保持菜的鮮綠顏色。
3. 沿春筍的長度，用刀把筍皮剔開，剝去筍皮，把根部老硬的部分切掉。筍尖空心的部分也同時去掉。用滾刀把筍切成塊，大火滾水汆燙 5 分鐘，撈出瀝水。
4. 百頁結用 1/2 茶匙小蘇打加約 500 毫升水泡 15 分鐘，拿出沖洗乾淨，同時擠出百頁結中心的水分，務求把所有小蘇打完全清除。
5. 把鹹肉、五花肉、薑和葱放在砂鍋內，加水到完全覆蓋肉面為止。大火加蓋煮滾，轉小火煮 90 分鐘。
6. 加入春筍，再煮 45 分鐘。
7. 薑和葱挾走不要，把鹹肉、五花腩取出切塊，和百頁結一同放回湯內，再煮 5 分鐘。
8. 最後放入小棠菜煮沸，試湯味調味，即成。

＊英文食譜見第 174 頁

材料

上海鹹肉	200 克
五花肉	200 克
薑	30 克
葱	2 根
小棠菜	150 克
春 / 冬筍	2 根
百頁結	200 克
小蘇打	1/2 茶匙

 烹調心得

1. 筍要預先汆水（出水），以去澀味。
2. 由於湯中有鹹肉，所以最後的調味要先試味才加鹽。

炒圈子

STIR-FRY PIG INTESTINE RINGS

　　炒圈子，又名紅燒圈子，圈子即豬大腸。炒圈子是上海市郊的傳統農家菜，可以說是真正的本幫菜。這個菜上桌時，豬腸呈一圈圈，所以叫做炒圈子。

　　口感香腴酥爛，不臊不膩，堂而皇之進入大飯店酒家的菜牌，歷久不衰。雖然叫做炒圈子，要大腸入味，必須是炒加紅燒。上海人做菜，炒和燒（炆）分不清，由清末民初叫到現在，上海人還是習慣叫做炒圈子，是本幫菜的代表作之一。

　　傳統的上海飯店做炒圈子，是用煸草頭來拌碟，是為了把豬腸提高身價，增加少許貴氣。沒有草頭的季節，可以用西蘭花或小棠菜，雖然沒有伴草頭的正宗，但這兩種菜都不會出水，別因伴碟的菜出水，而搞砸了一碟精心製作的圈子。

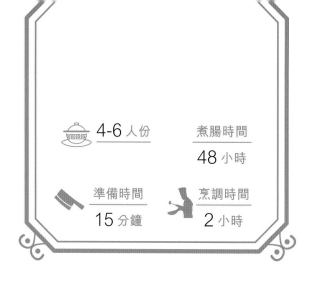

4-6 人份　　煮腸時間

48 小時

準備時間　　烹調時間

15 分鐘　　2 小時

材料

豬大腸	4 條
生粉	1 湯匙
鹽	1 湯匙
白醋	4 湯匙
薑片	30 克
白胡椒	1 湯匙（輾碎）
白糖	2 湯匙
薑絲	20 克
紹興酒	2 湯匙
紅燒醬油	2 湯匙
葱油（做法見第 10 頁）	1 湯匙

 烹調心得

選購豬大腸，首先必須要求新鮮，
應盡量在大清早去買，買回家立刻
清洗，如果宰殺後好幾個小時還未
處理，穢物在豬腸裏滋生細菌，豬
腸的臭味再怎樣洗也洗不掉。

做法

1. 豬大腸切去粗大的腸頭部分約 5 厘米不要，切去尾段一部分不要，只用約 50 厘米中間腸段。在水喉下把內腸翻出約 10 厘米，上下反轉，左手持翻出部分，右手在上面手持長長的大腸，用水沖向翻出的部分，水力令大腸徐徐滑入開口，整條大腸就反了內壁出來了。這時用生粉抓洗，用水沖淨，再用鹽抓洗，再沖淨，然後把大腸翻過來，再抓洗一次。每條大腸剪成兩段。

2. 用一鍋冷水，放入大腸、2 湯匙白醋、薑片和白胡椒碎，開大火至沸，煮 15 分鐘，轉中小火煮 1 小時，換水，加 2 湯匙白醋，再煮半小時至腸脸，撈出，用水沖洗乾淨，瀝乾，切成 1.5 厘米小段。

3. 用 2 湯匙白糖做炒糖色（見第 10 頁）。

4. 燒熱 1 湯匙油，爆香薑絲，放入大腸爆炒，加入紹興酒、紅燒醬油和糖色，大火炒至大腸稍為帶焦糖色，。

5. 加入葱油炒勻即成，伴以蔬菜盛碟。

※ 英文食譜見第 175 頁

品味上海菜 ❀ 家常菜

油麵筋塞肉 爁青菜

SAUTÉED WHEAT GLUTEN WITH VEGETABLES

這是一個道地的上海家常菜，用料經濟，味道鮮美。炮製此菜要花點心思和時間，需要把每一個油麵筋剪開一個口子，再把絞豬肉釀進油麵筋裏。

4 人份

準備時間　　　烹調時間
10 分鐘　　　10 分鐘

材料

絞豬肉	200 克
油麵筋	8 個
生抽	1 茶匙
糖	1 茶匙
生粉	1 茶匙
麻油	1 茶匙
葱	1 條（切段）
老抽	1 湯匙
鹽	1/4 茶匙
糖	1/2 茶匙
小棠菜（青江菜）	200 克

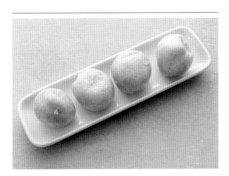

油麵筋

做法

1. 絞豬肉用刀再剁細，加 1/2 茶匙生抽、糖、生粉和 1 湯匙水拌勻，拌入 1 茶匙麻油。
2. 把每一個油麵筋剪開一個口子，再把絞豬肉釀進油麵筋裏。
3. 把小棠菜洗淨，用水焯 1 分鐘，撈起瀝乾水分。
4. 燒熱 2 湯匙油，爆香葱段，放入油麵筋煎香釀豬肉的一面，再放入 1/2 茶匙生抽、老抽、鹽、糖和 250 毫升水，煮沸，轉小火煮約 5 分鐘，放入菜再煮 5 分鐘，用生粉勾薄芡即成。

＊英文食譜見第 176 頁

炒蝦腰

STIR-FRY PRAWNS AND
PORK KIDNEY

「上有天堂，下有蘇杭」，如此美譽，當然一定離不開美食。杭州是魚米之鄉，風調雨順，她的小食多彩多姿，琳琅滿目。杭州人得天獨厚，性格是自得其樂，優哉悠哉。他們會一絲不苟地炒一小盤菜，或者熬一小鍋湯，繁複多樣，就是為了吃一碗心滿意足的好麵，也只有杭州人能有如此能耐，這種生活品味的神韻，如非杭州人，是無法心領神會的。如果你去到杭州，一定不容錯過吃一碗精心製作的湯麵，而杭州的傳統杭幫麵中，有一道蝦腰麵，就是用蝦和豬腰炒好來佐一大碗湯麵。上海人更絕，把蝦和豬腰炒成一道精美的小菜。

豬腰即豬的腎，有和理腎氣、利膀胱的功效。把豬腰切開，裏面的白色筋為腎上線，這就是羶味的來源。處理的方法是將豬腰開邊切開，用刀切去白筋和深紅色部分，洗淨，放在加了醋水中浸半小時，撈出後再用清水沖漂；在背部用斜刀剖成腰花，再用清水浸泡至無血水為止，中途換水 2 至 3 次，這就沒有羶味了。由於豬腰口感爽嫩，是各省菜式都喜歡用的食材，菜式例如：爆雙脆、燴腰片、醬爆腰花、木耳腰花湯。

4 人份		浸泡時間 30 分鐘
準備時間 15 分鐘		烹調時間 10 分鐘

材料

豬腰	1 個（約 150 克）
鮮蝦	300 克
白醋	1 湯匙
鹽	1 茶匙
醬油	1 湯匙
糖	1/2 茶匙
紹興酒	1/2 湯匙
生粉	1/2 茶匙
薑汁	1 湯匙
薑片	20 克
葱白	2 條（切段）
麻油	1/2 茶匙

 烹調心得

1. 購買時要注意，挑選有光澤、堅實、無萎縮的新鮮豬腰。
2. 把豬腰剳花是比較容易控制成熟度和更美觀。
3. 要為豬腰辟腥，用白醋浸泡是最有效的方法。

做法

1. 豬腰橫切成上下兩片（圖1-2），把白色的腰臊連同旁邊深紅色的組織全部切除（圖3），有深紅的部分也要挑出丟掉（圖4-5）。

2. 把豬腰泡在500毫升的清水內，加白醋，泡30分鐘，再用清水洗淨。

3. 在豬腰面上用斜刀剞腰花（圖6），剞痕相距約為3毫米，深度約為豬腰厚度的一半，再橫切成約3厘米寬的片，泡清水直至烹調前才取出，中途可換水2至3次。

4. 鮮蝦剝殼，去掉蝦頭，從背部剖開，挑出蝦線，用1茶匙鹽抓洗，再在水喉下沖淨。

5. 把醬油、糖、酒、生粉和1湯匙水混合成醬汁，下鑊前再拌勻。

6. 燒滾1公升水，加薑汁，放入豬腰，水再滾時立刻撈起，瀝乾。

7. 燒熱2湯匙油，爆香薑片和葱段，放入蝦，炒至半熟，加入豬腰和醬汁兜勻，大火收汁，拌入麻油即成。

＊英文食譜見第177頁

品味上海菜 ✿ 家常菜

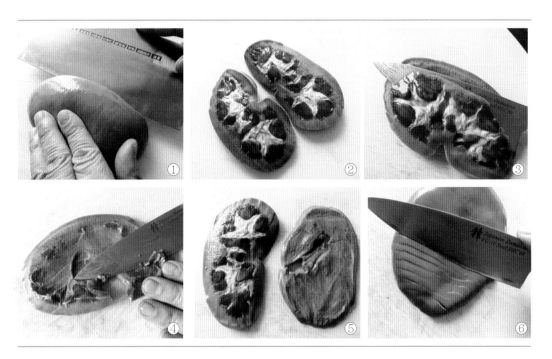

① ② ③

④ ⑤ ⑥

雞汁百頁包

TOFU ROLL
IN CHICKEN SOUP

　　有關豆腐的發明者，眾說紛紜，有說是西元前 164 年，淮南王劉安在八公山上採藥煉丹時，無意中以石膏點豆汁得來的靈感而發明的，也有考證說豆腐是在唐代或五代才有文獻的記載，所以應該把豆腐的發明年代推遲到唐末五代。不管怎樣說，中國是豆腐的發源地是沒有任何爭議的。用黃豆製成的食品很多，廣東人比較熟悉的是板豆腐、嫩豆腐、山水豆腐、腐竹、腐皮、豆腐泡等，而百頁（又名千張）是江浙地區喜愛的豆製食材。做法是將泡軟的黃豆加水磨成豆漿煮沸濾渣後，加凝固劑凝成「豆腐腦」，用布摺疊壓製成薄片狀。新鮮的百頁是白色的，可與其他食材直接烹煮。若是黃色的乾百頁，便要先經過處理才可入饌。用百頁做的代表性菜餚有：雪菜毛豆百頁、百頁結燒肉、雞汁百頁包等。

　　乾百頁（圖 1）可以在南貨舖買到，一般的包裝是一疊十張，用不完可以放在冰箱裏保存。百頁分兩種，一種是較薄的，可用來包餡做百頁包，另一種是較厚的，適合做百頁結。做百頁包在購買的時候請說明是要買薄的一種。

4 人份

準備時間
20 分鐘

烹調時間
15 分鐘

材料

乾百頁	5 張
絞豬肉	250 克
鮮蝦	50 克
娃娃菜	200 克
鹽	1/2 茶匙
糖	1/2 茶匙
生抽	1 茶匙
生粉	1 湯匙
清雞湯	500 毫升
梳打粉	1 茶匙

（亦稱食粉或小梳打）

| 青蒜 | 1 紮 |
| 胡椒粉 | 少許 |

烹調心得

1. 用小梳打開水浸百頁，可令百頁軟化及顏色變白（圖 2），但不要浸太長時間，否則會破壞百頁的組織。水的溫度會影響泡浸百頁的時間。

2. 煮百頁包容易散開，所以用蒜葉捆住，但也可以用牙籤把百頁包穿住，上桌前記得把牙籤取出。

3. 蒜葉見熱水即軟，燙過後要馬上拿出。

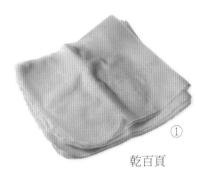
①
乾百頁

②
處理過的百頁

品味上海菜 家常菜

做法

1. 把梳打粉放進 1 公升溫水裏開勻,再把乾百頁放進,浸泡約 15 分鐘到顏色變白,再用清水徹底漂洗百頁,除去所有的梳打味。
2. 娃娃菜洗乾淨後切碎,放鍋裏用清水煮熟,撈起,放涼後用手把菜裏的水擠乾,再剁碎。
3. 鮮蝦去殼後稍為斬剁,不要過分細剁。
4. 把絞豬肉、鮮蝦、娃娃菜、鹽、糖、生抽、胡椒粉和生粉拌勻。
5. 把百頁一張分成四張,每一張放 1 湯匙肉餡,包成小包。
6. 把蒜葉切下,在開水裏一燙,馬上拿出,再用小刀順着蒜葉從中間剖開為二。
7. 每一小包用蒜葉輕輕捆住。
8. 把雞湯煮沸,百頁排好在鍋裏,煮沸後轉小火,煨 10 分鐘。可另加鹽調味。
9. 吃時用深碗裝好連湯上桌。

＊英文食譜見第 178 頁

生煸草頭 | STIR-FRY ALFALFA

　　生煸草頭是上海的特色名菜，草頭是苜蓿的一種，也俗稱三葉菜。苜蓿是在漢武帝的時代從西域引入。根據司馬遷《史記》第123卷〈大宛列傳〉所記載：「宛左右以蒲陶為酒，富人藏酒至萬餘石，久者數十歲不敗。俗嗜酒，馬嗜苜蓿。漢使取其實來，於是天子始種苜蓿、蒲陶肥饒地。及天馬多，外國使來眾，則離宮別觀旁盡種蒲萄、苜蓿極望。」文中並沒有說明漢使是誰，但是從記載中，這個漢使似乎不是張騫。另有一個說法是貳師將軍李廣利從大宛引入苜蓿。不管真情如何，苜蓿是在漢代從西域傳入中國，應該是沒有異議的。

　　苜蓿原來是作為養馬的牧草，營養豐富，多年生長，在北方種植一年可有一至兩次收成，在南方更可以收割三次。「一勞永逸」的成語就是來自苜蓿種植，因為種一次便可收成多年。除了作為飼料外，苜蓿的嫩芽也可以吃，在外國的超市，也有賣苜蓿芽，一般是用作沙律菜。在中國，歷史上也有很多有關苜蓿的飲食記載，上海人用來做菜，只選取苜蓿上的三片嫩葉，故稱之為草頭，是上海郊區農民的經濟作物。

　　生煸草頭是上海名菜，「生煸」亦即粵菜的「生炒」。生煸草頭要做得好，要講究草頭新鮮、火候、手勢，以及下油、鹽、酒、糖的次序和炒的速度，錯一就可能前功盡棄。草頭是帶香氣的野菜，不要浸水，以免浸去特有的香味。草頭容易熟，煸炒時要快手，要保持草頭的青翠綠色和青草味。至於這道菜應否放醬油，那是屬於烹調的流派和個人喜好，很難斷定對錯，而生煸草頭很吸油，用兩份豬油加一份生油來炒，味道極佳。

品味上海菜 · 家常菜

4 人份

準備時間
5 分鐘

烹調時間
1 分鐘

草頭，在上海南貨店有售。

材料

新鮮草頭	600 克
糖	1 茶匙
鹽	1/2 茶匙
生抽	1 茶匙
豬油	2 湯匙
生油	1 湯匙
薑	2 片
高粱酒	1 湯匙

 烹調心得

每次炒草頭的量不能太多，否則會因為 10 秒內難以炒勻而導至炒過火。

做法

1. 草頭稍為沖洗，不要浸水，用篩網裝着搖晃瀝乾，瀝乾時間要長一些，盡量少帶水分。

2. 糖、鹽和生抽用小碗拌勻成調味汁。

3. 大火燒熱豬油和生油，下薑片爆香後，加入草頭，快速急炒。

4. 炒的同時加入調味汁，共炒不要超過 10 秒，即用碟盛起，再立即上噴高粱酒。

＊英文食譜見第 179 頁

雪菜毛豆
百頁結

PRESERVED POTHERB
MUSTARD,
SOY BEANS AND
BAIYE KNOTS

上海菜可能給人一般的印象是濃油赤醬、口味較濃，其實上海菜也有清淡的一面。這道充滿豆香的雪菜毛豆百頁結，是當地的特色家常菜。

4 人份

準備時間　15 分鐘

烹調時間　10 分鐘

材料

百頁結	10 個
雪菜	200 克
毛豆仁	250 克
薑絲	10 克
糖	1/2 茶匙
清雞湯	60 毫升
食用小蘇打	1 茶匙

百頁結

做法

1. 把食用小蘇打粉放進 1 公升溫水裏開勻，再把百頁結放進，浸泡 15 分鐘到顏色變白，用清水徹底漂洗百頁結，除去所有的蘇打味，再汆水 3 分鐘。
2. 雪菜用清水浸泡 10 分鐘，擰乾，切碎。
3. 毛豆仁汆水 5 分鐘，瀝乾。
4. 燒熱 2 湯匙油，大火爆香薑絲，放入雪菜、毛豆和糖，炒約 1 分鐘，加雞湯和百頁結，煮至收汁即成。

＊英文食譜見第 179 頁

品味上海菜 ❀ 家常菜

豆瓣莧菜

AMARANTH GREENS WITH
BROAD BEANS

冷藏的蠶豆肉在南貨店四季都有售。烹調前，用水焯5分鐘，就可辟去雪藏的味道。

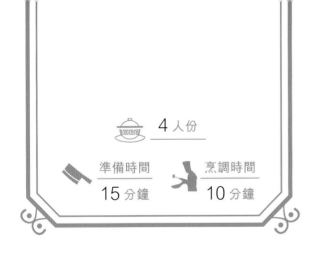

4人份

準備時間
15分鐘

烹調時間
10分鐘

材料

冷藏包裝蠶豆肉	150克
紅／綠莧菜	600克
蒜頭	2瓣（切片）
鹽	1/2茶匙

做法

1. 冷藏蠶豆肉用水焯5分鐘，瀝乾。
2. 莧菜洗淨，切掉菜根，再切成兩段。
3. 燒熱2湯匙油，爆香蒜片，下蠶豆和莧菜，大火炒至莧菜軟身，加鹽兜勻即成。

＊英文食譜見第180頁

品味上海菜 ❀ 家常菜

註：患蠶豆症（又稱「G6PD缺乏症」）人士絕對不宜進食此菜。

松子雞米

| CHICKEN WITH PINE NUTS

　　上海菜中的松子雞米，是源自川菜的「小煎雞米」，再結合上海人的口味而發展出來的。川菜中有稱為小煎小炒的烹調法，這是一種完全來自平民百姓的菜式，小煎小炒其實都是爆炒，要求急炒上菜，所以一是爐火旺油溫高，夠鑊氣；二是所有材料事先不過油（走油），一定是生炒；三是最後一定是大火埋芡迅速收汁，這就是川菜小煎小炒的特色。

　　由四川菜的「小煎雞米」到上海菜的「松子雞米」，少了四川泡椒和蒜苔，沒有了川味，加了炒香的松子，多了一份江南的婉約清秀；另一個它的姐妹作是雞米豌豆，也是上海的家常小菜。

　　松子仁含不飽和脂肪酸，有降低血脂，預防心血管病的功效；豐富的維他命 C，能軟化血管、延緩衰老，是中老年人理想的健康食物。

4 人份

準備時間
15 分鐘

烹調時間
10 分鐘

材料

雞腿肉	200 克
松子仁	50 克
鹽	1/2 茶匙
生粉	1 茶匙
上海特級醬油	1 茶匙
陳醋	1/2 茶匙
糖	1 茶匙
麻油	1/4 茶匙
薑米	1 湯匙
蒜頭	2 瓣（切粒）
紹興酒	1 茶匙

 烹調心得

把調味料預先拌勻做成碗芡，是
為了縮短翻炒的時間，令雞肉
更嫩。

做法

1. 雞腿肉挑去筋，洗淨，先用刀背
   拍鬆，再切成 1 厘米粗的雞粒，
   放入鹽和 1/2 茶匙生粉拌勻。

2. 把上海特級醬油、陳醋、糖、麻
   油、1/2 茶匙生粉和 1 湯匙水用
   小碗拌勻成調味碗芡，下鍋前再
   拌勻。

3. 松子仁用白鑊炒至酥香，備用。

4. 大火燒熱 2 湯匙油，放入薑米爆
   香，加雞肉和蒜頭粒，炒至雞肉
   變白約八成熟。

5. 灒紹興酒兜勻，加入松子仁同
   炒，最後把調味碗芡倒入，炒勻
   即成。

   ＊英文食譜見第 180 頁

小黃魚燒豆腐

BRAISED YELLOW CROAKER WITH TOFU

　　不少人有誤解，以為大黃花魚就是小黃魚長大了，其實小黃魚（學名 Larimichthys polyactis）不同於大黃花魚（學名 Larimichthys crocea），是同屬石首科的兩種魚，形態相似，生長習性相近。大黃花魚的中文名稱很多，各地有不同叫法，有大黃魚、金龍魚、紅瓜、黃金龍等，遍佈於中國黃海南部、東海和南海；小黃魚也叫做梅魚、小春魚、小黃瓜魚、厚鱗仔、花魚等名稱，分佈於中國的黃海、渤海和東海，以及南韓的水域。

　　野生的大黃花魚的尾柄收窄，全身呈金黃色，身長 40-60 厘米，甚至可長到更大。大黃花魚生長在較深海域，每年 4 至 6 月就會向淺海迴遊，夏天前產卵，然後在沿岸棲息，以魚蝦為食，到了秋天又遊回深海過冬。傳統江浙名菜，有蔥燒黃魚、松子黃魚、糖醋黃魚、雪菜大黃魚等。

　　小黃魚全身呈淺黃色，身長 20 厘米左右，是不會再長大的。小黃魚每年春季在沿岸短途迴遊，3 至 5 月間產卵，秋季再返回較深的海域。野生的小黃魚尾柄收窄細長，養殖的小黃魚尾柄較短。

　　大小黃魚的肉質都鮮嫩無比，千百年來深受中國人愛戴，曾經因長時期過度使用拖網圍捕，使黃魚幾乎瀕臨絕種。1985 年福建官井洋建成大黃魚國家級水產種質資源保護區，為野生黃花魚的棲息及產卵提供保護。近年更建成高科技環保的大型深海設備，放養黃花魚及小黃魚，環境盡量近似野生，從此市場上多了不少供應，大家有口福了。

品味上海菜 · 家常菜

4 人份

準備時間
20 分鐘

烹調時間
15 分鐘

材料

小黃魚	3 至 4 條
（約 450 克）	
豆腐	250 克
鹽	1 茶匙
薑片	10 克
蔥	1 條（切段）
上海特級醬油	2 湯匙
糖	1/2 茶匙
紹興酒	1 湯匙
鎮江醋	1 茶匙
生粉	1 湯匙

①

②

烹調心得

黃花魚體內有個充氣的魚鰾，所以魚肚部分肉比較薄，如果開膛取內臟的話，魚肚部分在煎炸時很容易弄碎。黃花魚內臟不多，用筷子旋轉抽出較為容易，能保持魚身的完整；注意先要在肚下切一小刀口，否則就比較難一次過抽出內臟和魚鰾。

做法

1. 小黃魚去鱗去鰓，不開肚，在肛門位置橫切一刀口（圖 1），再用兩根筷子從魚頭中插入（圖 2），夾緊魚鰾和內臟向上旋轉抽出，再把魚身和肚內沖洗乾淨，瀝乾。
2. 小黃魚兩邊抹上一層薄薄的生粉，用油煎至金黃，取出備用。
3. 豆腐切厚片，用 500 毫升溫水加鹽泡 15 分鐘，瀝乾。把豆腐略煎至兩面金黃，取出備用。
4. 把醬油、糖、鎮江醋、生粉，加 250 毫升水拌勻成碗芡。
5. 砂鍋中燒熱 2 湯匙油，爆香薑蔥，放入魚和豆腐排好，在鍋邊潷酒，倒入碗芡沸煮 1 至 2 分鐘，即成。

＊英文食譜見第 181 頁

品味上海菜 ✿ 家常菜

毛豆蝦仁

這是一道賞心悅目的菜式。微粉紅的蝦仁，配上翠綠的毛豆，煞是美麗。

SHRIMPS WITH PODDED
SOY BEANS

4 人份

準備時間 10 分鐘　　烹調時間 10 分鐘

毛豆

材料

河蝦仁	300 克
毛豆仁	150 克
鹽	1/4 茶匙
雞蛋白	1/2 個
生粉	1 茶匙
薑片	10 克
紹興酒	1 湯匙

烹調心得

解凍冷藏蝦仁，最好的方法是把蝦仁泡在冰水裏，讓蝦仁慢慢均勻地解凍，蝦仁肉就不會變黴。冷藏的河蝦仁在香港的南貨店有售，也可以用南美白蝦仁代替。

做法

1. 河蝦仁洗淨，瀝乾，放在碗內，加入 1/4 茶匙鹽，用手拌至起膠，加入雞蛋白拌勻，再拌入生粉。

2. 燒熱一鍋水，把毛豆仁沸煮 5、6 分鐘，瀝乾。

3. 在鑊裏把 500 毫升油熱至低溫（約 120℃），放入蝦仁，迅速用筷子划散（約 15 秒），立即取出，瀝油。把鑊裏的油倒出，只留 1 湯匙油。

4. 大火把油燒熱，爆香薑片，放入毛豆仁，再倒入蝦仁爆炒至蝦仁全熟，沿鍋邊潷酒，快炒至酒完全蒸發，即可上碟。

＊英文食譜見第 182 頁

紅燒下巴甩水

BRAISED FISH HEAD
AND TAIL

　　上海人愛吃青魚，青魚即廣東和香港的黑鯇，分別之處，上海的青魚是吃螺螄長大的，廣東的黑鯇是吃草長大的，所以叫做草魚。青魚的味道較為鮮美，肉質嫩滑而刺少，沒有土腥味。冬天的青魚最肥美，到上海去，一定要吃青魚。

　　清代末年，上海和無錫等地已經盛行吃青魚，最著名的是始於清同治年間的老正興菜館，號稱「鮮活大王」，以青魚為材料，烹調出上海風味的菜式，有紅燒下巴、扇形甩水、湯卷（魚頭魚腸魚尾）、青魚煎糟、青魚肚襠（魚腩）、青魚禿肺（魚肝）、燒頭尾等，都是本幫名菜。

　　下巴，即魚頭連下顎，廣東人叫魚下顎做魚骹。魚頭的面頰肉、眼瞳和下顎位置的魚肉，是整條魚肉質最細嫩的位置；甩水，又叫做划水，就是魚尾，魚尾活動得多，雖然多刺，但魚肉非常嫩滑。以上海紅燒醬油做的紅燒下巴甩水，肥美嫩滑，味道香濃而不肥膩。在買不到青魚的地方，做這個菜可用鯇魚代替。

4 人份

準備時間
10 分鐘

烹調時間
20 分鐘

材料

魚頭連下巴	1 個
魚尾	1 條
鹽	1/2 湯匙
生粉	3 湯匙
葱	2 條（切段）
紹興酒	2 湯匙
薑汁	1 湯匙
紅燒醬油	2 湯匙
糖	2 湯匙

做法

1. 魚頭連下巴刮鱗，斬開兩邊，去鰓，洗淨瀝乾。魚尾刮鱗，洗淨，垂直連切成三條。用鹽把魚頭魚尾拌勻，醃 15 分鐘，再用 2 湯匙生粉拌勻。

2. 燒熱 250 毫升油，分別放入魚頭和魚尾，煎香兩邊，取出。

3. 鍋內留 2 湯匙油，爆香葱段，放入魚頭和魚尾，灒酒，加入薑汁、醬油、糖和 250 毫升水，煮沸，轉小火，加蓋焖約 5 分鐘至入味，掀蓋，大火收汁。

4. 把魚鏟出裝盤，把鍋內的汁用 1 湯匙生粉加水勾濃芡，淋在魚上即成。

＊英文食譜見第 182 頁

響油鱔糊

| SAUTÉED EEL

　　黃鱔，又稱長魚，是中國自古以來已經有的淡水魚類，早在兩千多年前的《山海經》早有記載。黃鱔營養豐富，藥用價值高，中醫認為吃黃鱔有補腦、健身、祛風通絡的功效，被很多古代的醫書記載收藏。據說每年端午之後 1 個月的黃鱔最滋補，民間有「小暑黃鱔賽人參」之說。

　　響油鱔糊，是上海的傳統風味小菜，響油是形容上枱時蒜油吱吱作響。清朝道光年間，淮揚地區流行「清炒軟兜」的菜式，據說是源於安徽菜。軟兜即黃鱔，當地用的是野生小黃鱔，每一條只能剖出兩條鱔肉，每一段約 16 至 17 厘米長。清炒軟兜便是油泡後清炒，配料用的是蒜片、薑絲、紹興酒、胡椒粉，肉脆而皮滑，一口吃一條。野生小黃鱔在廣東和香港不易買到，也有季節所限，一般是用黃鱔切絲做炒鱔糊。不同的是，上海的小黃鱔是用灼水熟殺，然後再剖絲的處理方法；而在湖南、廣東和港澳地區，吃黃鱔一定是吃活宰。

　　購買黃鱔時，要注意體型平均，盡量不要買那些從上半身到肛門的一段特別肥大的黃鱔，因為這種黃鱔有可能經過藥物催促生長。買黃鱔時，一定要請店舖代為活宰及去骨。

4 人份

準備時間　15 分鐘

烹調時間　10 分鐘

 材料

黃鱔	500 克
生粉	1.5 湯匙
鹽	1 茶匙
上海特級醬油	2 湯匙
糖	1 茶匙
豬油	1.5 湯匙
薑絲	1 湯匙
紹興酒	1 茶匙
韭黃	100 克（切段）
胡椒粉	1/2 茶匙
蒜茸	2 湯匙
麻油	1 湯匙

 烹調心得

1. 爆的意思是要火猛手快，不要加蓋，開始下鍋炒前，預先用小碗把所有的調料準備好，炒的時候不會浪費時間，鱔條炒得過熟就會失去軟滑爽口的效果。

2. 韭黃見火易熟，不必擔心韭黃不熟，如果過熟就會韌。

 做法

1. 黃鱔去頭去骨，用 1 湯匙生粉和 1/2 茶匙鹽抓洗，用水洗乾淨後瀝乾，橫切成每段 8-10 厘米長，再直切成約 1 厘米寬的條狀。

2. 大火煮水大沸騰後熄火，倒入鱔條，用筷子迅速攪動，不要開火，目的在汆去血水，10 秒鐘左右撈起，用清水沖洗後瀝乾備用。

3. 把 1/2 茶匙鹽、1/2 湯匙生粉、醬油、糖和 1 湯匙水用一個小碗裝好拌勻，下鍋前再拌勻。

4. 大火燒熱豬油，放入薑絲和鱔條一起爆炒到鱔條乾身，然後在鍋邊潷入紹興酒，爆炒十多下後加入拌好的醬料，快手爆炒到鱔條熟透，放入韭黃段炒勻，再加少許胡椒粉兜勻，裝盤。

5. 用筷子把鱔條撥開，中間留一小洞，在洞中放入蒜茸。

6. 用另外一隻乾淨鑊燒沸 1 湯匙生油和 1 湯匙麻油，淋在蒜茸堆上，立刻上菜。要求在上菜時，蒜油仍吱吱作響。

＊英文食譜見第 183 頁

品味上海菜 家常菜

油醬毛蟹

SAUTÉED RIVER CRAB
IN SOY SAUCE

毛蟹即大閘蟹，油醬毛蟹是上海傳統名菜，也叫做麵拖蟹。選用六月當造的毛蟹（六月黃），加上剛剛上市的新鮮毛豆，炒煮而成。六月黃毛蟹體型較小，但肥壯而蟹黃豐滿，裹上醬汁和豬油，加上翠綠的毛豆，是一道美味又亮麗的蟹饌。

 4-6 人份

準備時間　20 分鐘

烹調時間　15 分鐘

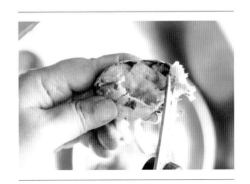

材料

毛蟹（小）	4 隻
毛豆仁	150 克
麵粉	2 湯匙
生粉	1 湯匙
葱	1 條（切葱花）
薑	20 克（切末）
紹興酒	2 湯匙
糖	1 湯匙
上海特級醬油	2 湯匙
生粉	1 茶匙（勾芡用）
豬油	1 湯匙

毛蟹（大閘蟹）

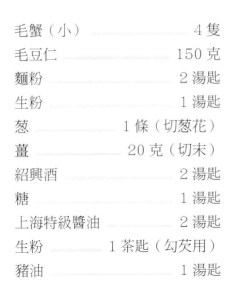
做法

1. 把每一隻毛蟹的腹部掀起丟掉，再從中間砍成兩半，去掉蟹鰓和蟹胃（見圖），洗淨。
2. 毛豆仁汆水，瀝乾。
3. 把麵粉和生粉拌勻，再把毛蟹切面蘸上粉。燒熱 2 湯匙油，把毛蟹切面煎至微黃，取出。
4. 鑊裏燒熱 1 湯匙油，爆香薑葱，加入毛蟹兜勻，灒酒，加毛豆仁、糖、醬油和 250 毫升水，煮至毛蟹全熟，用生粉勾芡，加入豬油兜勻即成。

＊英文食譜見第 184 頁

品味上海菜 ❀ 家常菜

墨魚大燶

STEWED PORK BELLY WITH CUTTLEFISH

上海人叫墨魚做目魚，所以這個菜也叫做目魚大燶，或者目魚紅燒肉，通常都會加上幾個雞蛋，不會浪費那香濃美味的紅燒汁。

4-6 人份

準備時間 20 分鐘　　烹調時間 90 分鐘

材料

五花腩	600 克
新鮮大墨魚	300 克
雞蛋	4 個
白糖	2 湯匙（糖色）
薑片	50 克
上海豆瓣醬	1/2 湯匙
紹興酒	3 湯匙
八角	2 粒
紅燒醬油	2 湯匙
鹽	1 茶匙
冰糖	40 克

做法

1. 五花腩出水切成 3 厘米厚塊，墨魚洗淨切成 4 厘米方塊。
2. 雞蛋用水煮熟，浸冷水，剝殼。用小刀每隻蛋直剞 5 至 6 刀，刀痕深至觸及雞黃。
3. 炒糖色（本書第 10 頁），備用。
4. 燒熱 2 湯匙油，爆香薑片和豆瓣醬，放入豬肉大火兜勻，加入紹興酒、八角和煮好的糖色，再加水至覆蓋豬肉，煮滾，收慢火加蓋燜 1 小時。
5. 放入醬油、鹽、冰糖和墨魚，燜 15 分鐘，放入雞蛋同煮 5 分鐘，改大火，收汁即成。

*英文食譜見第 185 頁

品味上海菜 家常菜

銀魚跑蛋

SCRAMBLED EGGS WITH NOODLE FISH

　　銀魚是一種常見的半透明白色小魚，生長在近岸海域，也生長在湖泊，江蘇省太湖是盛產銀魚的地方。銀魚跑蛋是上海家常小菜，雖然材料同樣是銀魚和蛋，但它與粵菜的白飯魚煎蛋或銀魚烙的做法不同，可以說是「滑蛋溜銀魚」，蛋少銀魚多，蛋醬稀滑地抱着銀魚，吃的就是嫩滑和鮮味。銀魚跑蛋是平凡中見技巧，用猛火燒紅鍋，全程急炒不過 10 秒就要盛起，要求炒出來的雞蛋不結塊，也不能過稀。為了形容炒的過程要手急眼快，所以叫做「跑蛋」。

🍲 4 人份

準備時間
10 分鐘

烹調時間
5 分鐘

材料

銀魚	350 克
雞蛋	2 個
葱	1 條（切絲）
鹽	1/4 茶匙
胡椒粉	1/8 茶匙
豬油	2 湯匙

銀魚

材料選購：

香港市場上有冷藏的盒裝太湖
銀魚出售。

做法

1. 銀魚洗淨，汆水 20 秒，撈起，
   用廚紙吸乾水分。

2. 雞蛋打勻，放入銀魚、葱絲、鹽
   和胡椒粉，拌勻。

3. 大火燒熱 2 湯匙豬油，倒入銀魚
   蛋漿，用筷子急炒 8 至 10 秒至
   蛋液凝結，即可裝盤。

＊英文食譜見第 185 頁

醬爆茄子

SAUTÉED EGGPLANT WITH THICK SAUCE

這是個深受歡迎的上海素菜，茄子吸滿了醬汁，軟糯可口，老少咸宜。上海豆瓣醬，味道鹹鮮而不辣，是做上海菜常用的醬料。

4 人份

準備時間 5 分鐘　烹調時間 5 分鐘

材料

茄子	600 克
鹽	1 茶匙
上海豆瓣醬	2 湯匙
糖	1 湯匙
薑末	2 湯匙
葱	1 條（切段）
葱	1 條（切葱花）
蒜蓉	1 湯匙
紹興酒	2 湯匙

上海豆瓣醬

做法

1. 茄子洗淨，用滾刀切成塊，泡在鹽水中，下鍋前瀝乾。
2. 豆瓣醬加糖和 2 湯匙水，拌成醬汁。
3. 燒熱 3 湯匙油，爆香薑末、葱段和蒜蓉，放入茄子，大火煸炒至茄子肉全熟，顏色變焦黃，灒酒，加入醬汁，煮至醬稠，掛在茄塊上，兜勻，撒上葱花即成。

＊英文食譜見第 186 頁

品味上海菜 ❀ 家常菜

雙菇麵筋

BRAISED WHEAT
GLUTEN WITH
MUSHROOMS

油麵筋的可塑性高，可與各種葷素菜搭配。這道菜內的麵筋又軟又糯，吸飽了冬菇和蘑菇的鮮味，非常可口。

4 人份

準備時間 5 分鐘

烹調時間 7 分鐘

材料

油麵筋	10 個
冬菇	4 朵
小蘑菇	8 個
葱白	2 條（切段）
糖	1 茶匙
生抽	1 茶匙
老抽	1/2 湯匙
生粉	1 湯匙（勾芡）
麻油	1 茶匙

 做法

1. 冬菇洗淨，用冷水泡軟，去蒂，切成兩邊。蘑菇洗淨。

2. 燒熱 2 湯匙油，爆香葱段，加入冬菇和蘑菇同炒 1 分鐘，加水至覆蓋食材，煮沸，放入油麵筋、糖、生抽和老抽，轉小火，加蓋燜 5 分鐘。大火收汁，用生粉勾芡，拌入麻油即成。

＊英文食譜見第 186 頁

糟溜魚片 | FISH IN DISTILLED GRAIN SAUCE

糟溜魚片，源自山東福山的魯菜，後來成為受歡迎的上海菜。相傳明朝隆慶年間，兵部尚書郭忠皋回山東老家福山探親，回京時帶同一位福山廚師。這位廚師做的糟溜魚片，享譽京華，皇室和滿朝文武都大為稱讚，幾年後廚師告老還鄉。一日，皇帝不思飲食，懷念那福山廚師做的糟溜魚片，皇后娘娘便派半副鑾駕，趕往福山降旨，將廚師的兩位徒弟召入宮為皇帝做糟溜魚片。這位福山廚師的家鄉，被後人稱為「鑾駕莊」，至今仍在。

香港的上海菜館和北京菜館都有做糟溜魚片，很受食客歡迎，幾十年來歷久不衰。「溜」這個做法，使人聯想到「蹓」，當然表示的是快手快腳之意。「溜」就是先把食材嫩油泡熟，或水灼熟，同時另起鍋煮好各種芡汁，把食材快速倒入芡汁中一拌即起，或者乾脆是把芡汁淋上食材上即完成。常見的溜煮法有滑溜、醋溜、軟溜、糟溜等，味道各有不同，但都同樣是用大火快速兜煮芡汁而成，目的是要求材料保持鮮嫩軟滑的特色。

糟溜，就是在用來「溜」魚片的芡汁中，加入了香糟，突出糟的醇厚濃香。我們介紹的是比較簡單的做法，用糟滷加紹興酒來代替香糟，難度不大，而瓶裝糟滷比較容易買到。

4 人份

準備時間
20 分鐘

烹調時間
10 分鐘

材料

魚肉	250 克
鹽	1/2 茶匙
雞蛋白	1 個
生粉	3 湯匙
乾木耳	5 克
筍片	50 克
薑片	10 克
糟滷	4 湯匙
紹興酒	1 湯匙
糖	2 茶匙
葱油（做法見第 10 頁）	2 湯匙

材料選購：

石斑魚、鱈魚、比目魚、青衣、桂花魚等的魚肉，都適合做糟溜魚片。

做法

1. 魚肉用水沖洗乾淨，斜刀切成約 3 毫米厚的魚片，下鹽，用手輕輕攪拌至起黏性，加入雞蛋白拌匀，再拌入 1 湯匙生粉，醃 15 分鐘。

2. 乾木耳浸發後撕成小塊，用水灼熟。筍片汆水。

3. 燒滾一鑊水，把魚片分別攤開，放入水中，熄火，讓熱水把魚燙熟，撈出。

4. 燒熱 1 湯匙葱油，爆香薑片，放入筍片、木耳、糟滷、紹興酒，糖和 250 毫升水一起煮沸，用 2 湯匙生粉開水勾成濃芡汁，然後放入魚片，淋上 1 湯匙葱油，輕輕一拌即起。

＊英文食譜見第 187 頁

宴客菜

世間榮貴月中人。嘉慶在今辰。蘭堂簾幕高卷，清唱遏行雲。持玉盞，斂紅巾。祝千春。榴花壽酒，金鴨爐香，歲歲長新。

—— 宋●晏殊《訴衷情》

菜式來自大戶人家的私房家宴。

這章節介紹的菜式如炸烹大蝦、水晶蝦仁、魚唇紅燒肉、楓涇丁蹄、紅燒河鰻、清蒸甲魚、蝦籽大烏參拌蔥黃、蟹粉豆腐等等，

看似步驟繁複而需要高超的烹飪技巧，其實只要細心和耐心，便人人都可以做到，絕對難不倒你。

水晶蝦仁

CRYSTAL SHRIMPS

水晶蝦仁，看似白玉，晶瑩剔透。一般廚師做水晶蝦仁，由於出菜要快，多數是採用粵式的滑炒。傳統做水晶蝦仁的方法，是採用了粵菜中的「中溫油泡」技術，例如粵菜中的「油浸筍殼魚」。把蝦仁用鹽糖入味後，上蛋白粉漿，在中溫熱油中靜靜浸泡至蝦仁僅熟，就能做到亮如水晶，鮮嫩爽口，一定不會過火，比炒蝦仁更為講究。

4 人份

準備時間
10 分鐘

烹調時間
4 分鐘

 材料

河蝦仁	300 克
鹽	1/2 茶匙
糖	1/2 茶匙
雞蛋白	1 個
生粉	1 湯匙

 烹調心得

1. 解凍冷藏蝦仁，最好的方法是把蝦仁泡在冰水裏，讓蝦仁慢慢均勻地解凍，蝦仁肉就不會變黴。

2. 冷藏的包裝河蝦仁，在南貨店有售；也可以用冷藏的南美白蝦仁代替，浸冰水解凍後一切開二，用來做水晶蝦仁，效果也很好。

 做法

1. 把河蝦仁用 1/4 茶匙鹽抓洗，用水沖約 5 分鐘，瀝水，再用廚紙吸乾水分。

2. 用餘下的鹽和糖把河蝦仁拌勻，加入雞蛋白，再拌入生粉。把蝦仁放入笓箕內，瀝去多餘的蛋漿。

3. 燒熱約 500 毫升的油至中溫（約 140℃，插入筷子見有少許泡冒出），放入河蝦仁，用筷子稍為撥散，熄火，用溫油把河蝦仁泡 2 分鐘至熟，撈出瀝油，即可裝盤。

＊英文食譜見第 187 頁

炸烹大蝦

DEEP FRIED JUMBO
PRAWNS WITH
PUNGENT SAUCE

幾十年前，這道上海菜炸烹大蝦，用的是東海的大對蝦，但現在對蝦已絕跡，改用大海蝦代替，也有餐館叫這道菜做「炸烹蝦段」或「炸烹明蝦」，名稱取決於蝦隻貨源的大小。「炸烹大蝦」在以小菜為主的上海菜中，增添了一道豪華的宴客大菜。

4-6 人份

準備時間 **15 分鐘**　　烹調時間 **15 分鐘**

 材料

大海蝦	600 克
清雞湯	2 湯匙
生抽	1 湯匙
鹽	1/2 茶匙
糖	1 茶匙
鎮江醋	1 湯匙
生粉	1/2 茶匙
紹興酒	1 湯匙
蒜頭	2 瓣（切片）
薑絲	10 克
葱	2 條（切絲）

 烹調心得

不要把蝦炸至全熟，七、八成熟即可，否則再煮的時候會把蝦煮得過火，蝦肉變韌。

 做法

1. 用牙籤把蝦從背部第二節挑出蝦腸，再用廚剪剪去蝦鬚、蝦腳。

2. 從蝦頭以下，沿蝦腹把蝦殼剪開，直到蝦尾。

3. 用小刀從尾部開始，沿已經剪開的蝦殼從尾到頭把蝦肉剖開，再拍上一層薄生粉。

4. 把生抽、鹽、糖、鎮江醋、生粉和 2 湯匙清雞湯，拌勻成醬汁碗芡。

5. 在鑊裏把 500 毫升炸油燒熱到中溫，把蝦炸至八成熟，取出，把炸油倒起，鑊裏只留 2 湯匙油。

6. 大火爆香蒜片和薑絲，放下大蝦，讚紹興酒兜勻，加入碗芡煮沸，翻炒至收汁，加葱絲，即成。

＊英文食譜見第 188 頁

品味上海菜 🏵 宴客菜

大烏參／豬婆參

　　常見的海參品種，以刺參、禿參、石參和烏參為主。中菜常用的大海參，俗稱豬婆參，肉厚肥胖，表面有灰質。海味市場上的豬婆參，分屬不同品種的烏參和石參。烏參科中的烏圓參（烏元參），體型短而渾圓飽滿，口感軟滑，因為供產較少，價格相對比較貴，其中產自印尼的烏圓參最受歡迎，產自中東的烏圓參則海水鹹味較濃；另外一種是屬於石參科，有白石參（白豬婆參）產自斐濟群島，口感軟滑，是市場上最被普遍採用的豬婆參品種；烏石參（黑豬婆參），口感爽脆，產自菲律賓南部海域；斑點石參（斑點豬婆參），外型有斑點，浸發後體型渾圓，口感爽中帶滑，產自毛里求斯。

　　挑選豬婆參，以外形肥大飽滿、乾度十足、不黏手的為靚貨。海參的支數代表參體的斤両，例如 1 斤有兩支，叫做兩頭，還有 3 頭和 4 頭等，代表了海參的大小。豬婆參的品種，可以按不同的菜餚來選購，例如做百花釀大海參或本書介紹的蝦籽大烏參，可選用石參科的白豬婆參或斑點豬婆參，口感軟滑；如果做原條海參上菜，建議選用烏圓參，成菜後海參賣相立體飽滿，口感較為爽滑，在台灣菜中很受歡迎。但是，中國人吃海參，有一個不成文的習慣，北方人喜歡吃口感軟爛的海參，廣東人喜歡吃較為爽滑的海參；所以，可按照自己的愛好來決定選用哪種豬婆參。當然，海參的軟和硬，是可以用浸發的時間來控制，浸發的經驗很重要。

乾白石參

已浸發的白石參

已浸發的斑點石參和乾斑點石參

豬婆參 的 浸 發 方 法

晚上把豬婆參放入一個潔淨無油可烹調的容器內，清水浸海參過面，開火煲 1 小時，收火焗至第二天早上。第二天早上換清水再煲 1 小時，又收火焗至晚上，再換清水煲 1 小時，收火焗至第三天早上。經過三次煲、浸、焗的豬婆參，體積已膨脹很多，可把表面灰泥盡量擦掉。剩下部分擦不掉灰泥的豬婆參，可瀝乾水後放入冰格 48 小時，再浸清水，基於冷縮熱脹的原理，灰泥便容易擦去。

這時用手輕輕捏一下，如果感覺豬婆參仍硬身，即未發透，就再用水煲 30 分鐘，之後焗 3 小時。然後再換清水，把豬婆參存放雪櫃低溫格再浸兩天。把徹底發透的豬婆參，掏出腸臟沙粒，沖洗後包好，放入冰格儲存，建議每次一起浸發兩三隻，煮食前解凍即可用。這種方法適合家庭廚房採用，全過程大約需要 5 至 6 天，但這方法不會把參皮剝去，可保海參的完整。

餐館酒樓浸發豬婆參，為了要節省時間，會在浸發前，把乾的豬婆參放在火上燒，灰質會連同海參皮一起剝落，全參露出海參肉，這樣浸發三四天就可以用了。這個方法雖然時間較短，但不建議家庭主婦採用，一是用火燒要注意安全，操作必須有經驗；二是萬一過火燒焦，海參就報廢了。

蝦籽大烏參拌葱黃

BRAISED SEA CUCUMBER WITH SHRIMP ROE AND CHINESE LEEK

上世紀 30 年代，上海的十六浦是重要的港口之一，而十六浦集中了海味乾貨的店舖，專營由各國進口的名貴海味，是上海最大的海味批發零售市場，地位等同今日香港上環的海味街。

上海人一向喜歡吃刺參，不熟悉怎樣浸發大烏參。位於十六浦的一家名店「德興館」的楊姓主廚，有感於大烏參滯銷，於是反覆研究浸發的方法，又利用蝦籽來增加鮮味，結果一舉成名；蝦籽大烏參成為上海的名菜，歷久不衰，流傳至今。

葱黃，就是大葱的芯，撕去兩三層葱衣取嫩嫩的葱黃，含有揮發油產生的特殊香味，煎過的葱黃，口感柔軟，減少了大葱的辣性，多了一份甜味。這是山東魯菜的講究吃法，我們被啟發自在濟南周村吃過的商埠菜「黃葱燒海參」，從此愛上了葱黃。用煎軟的葱黃來伴蝦籽大烏參，嫩滑的葱黃吸收了汁醬和蝦籽的味道，香味濃郁，但又不會喧賓奪主，令到整個菜的味道更醇厚，效果之好，非一般拌碟的青菜能及。

6-8 人份

準備時間　　　　烹調時間
1 小時　　　　　15 分鐘

材料

大烏參（豬婆參）	1 條
清雞湯	250 毫升
蝦籽	2 湯匙
大葱（京葱）	4 條
葱油（做法見第 10 頁）	4 湯匙
薑片	30 克
蒜頭	2 瓣（切片）
紹興酒	2 湯匙
糖	1 湯匙
鹽	1/2 茶匙
老抽	1 湯匙
生粉	適量（勾芡用）

做法

1. 用 250 毫升清雞湯把已發好的大烏參慢火煨煮 30 分鐘到 45 分鐘，至海參變軟身夠腍。

2. 用白鑊慢火把蝦籽炒香，盛起備用。

3. 大葱撕走兩三層外衣，只取葱黃（即葱芯），切成 4 厘米段，用 3 湯匙葱油慢火煎炸至微黃，取出。

4. 燒熱鑊中餘油，爆香薑片和蒜片，轉中火，放入蝦籽煸炒至香，潷酒。

5. 加入烏參、鹽、糖、老抽和煨烏參的雞湯，煮沸，放入葱黃，轉小火煨至濃稠，勾芡，最後淋上 1 湯匙葱油，即可上碟。

＊英文食譜見第 191 頁

品味上海菜 ✿ 宴客菜

蟹粉豆腐

TOFU WITH CRAB MEAT AND CRAB ROE

　　蟹粉豆腐，是上海和江蘇省的經典名菜。長江中游的湖泊中盛產清水大閘蟹，其中以江蘇的陽澄湖和太湖的大閘蟹最為著名。每年秋天是江南蟹季，養殖和銷售大閘蟹的蟹農和商家，除了出售整隻活生生的大閘蟹之外，都會把部分體型較小的大閘蟹蒸熟，拆出蟹肉和蟹黃混合起來以盒裝出售，這便是蟹粉。

　　蟹粉的味道極為鮮甜，配搭麵食、菇類、蔬菜和豆腐，成為一道道嫩滑鮮美的蟹粉菜餚，蟹粉湯包更是馳名中外。小籠湯包最早出現於淮安和揚州，傳到了上海，外形變得小巧，也就是香港人常見的小籠包，蟹粉小籠包更是秋天的美食。

　　蟹粉和豆腐是絕配，是江南蟹粉菜系列中最受歡迎的菜式之一，色澤鮮艷，華貴奪目，是請客吃飯的好菜式。為了突顯這道菜的矜貴，我們在煮好的蟹粉豆腐上，再放上一些紅色的新鮮蟹籽，使賣相更加亮麗，而蟹籽也令這道菜的味道和口感，更上一個層次。

4-6 人份

準備時間 10 分鐘

烹調時間 5 分鐘

材料

鹹蛋	3 個
薑	50 克
紅蘿蔔	50 克
豆腐	2 塊
鹽、糖	各 1 茶匙
鎮江醋	1 茶匙
生粉	2 茶匙
蟹粉	3 湯匙
蟹籽	200 克

 烹調心得

1. 鹹蛋黃分兩次加入，第一次（做法 3）是取其味，第二次（做法 5）是取其色。
2. 鎮江醋的作用是提鮮蟹味，但是不能多放，多放就會影響蟹粉的顏色。

 做法

1. 把鹹蛋蒸熟，蛋黃壓碎，蛋白不要。
2. 薑切薑米，紅蘿蔔刨蓉，豆腐切 1.5 厘米豆腐丁。
3. 在鑊裏用中火燒熱 2 湯匙油，先把薑米和紅蘿蔔蓉爆香，放入 3/4 份量的鹹蛋黃和 60 毫升水，煮沸。加入鹽、糖和醋調味。
4. 把豆腐的水分瀝乾，加進鑊裏，用鏟輕輕兜勻，略煮約半分鐘，再用生粉勾薄芡，盛在深盤中。
5. 在鑊中放 2 至 3 湯匙水，煮沸，加入蟹粉和餘下的鹹蛋黃，拌勻，盛出淋在豆腐上。
6. 最後把蟹籽撒在蟹粉上，即成。

※ 英文食譜見第 189 頁

清蒸甲魚

STEAMED SOFT SHELL
TURTLE

這是一道在民國時期，由當時上海的粵菜餐館傳入，之後成為上海本地菜式之一。甲魚，廣東人稱為水魚，古稱為鱉。甲魚生長在河流湖泊中，以小魚小蝦為食，其分佈很廣，長江和珠江流域、雲南、貴州、廣西、海南島都有出產甲魚。以前吃甲魚，是貴價的菜式，十多年前，中國發明了利用溫泉水使甲魚避開冬眠，而令其加快生長，逐漸培養成養殖甲魚的技術，使甲魚的市場價格下降，甲魚菜式得以走入平常百姓家。

甲魚味道鮮美，營養豐富，中醫認為吃甲魚能補中益氣，治風濕痹痛，滋補有益。甲魚最佳部分是四周下垂的水魚裙，味鮮，肉軟而腴，古人讚稱為「肉加十臠尤難比」。甚至連僧人也為之垂涎，見《五代史補》中曰：「南唐僧人謙光嗜鱉，國主戒之。對曰：老僧無他願，但得鵝生四腿（掌），鱉長兩群（裙）足矣，國主大笑。」

要注意的是，甲魚之所以味道鮮美，是因為肉中含有較多的組胺酸，但自然死後組胺酸會被細菌分解而產生有毒物質，所以購買甲魚切記必須要活宰。甲魚分野生和養殖兩種，野生的腹部呈黃色，而養殖的呈白色，很容易分辨。

4 人份

準備時間
30 分鐘

烹調時間
30 分鐘

 做法 . . . .

材料

甲魚 / 水魚	1 隻
冬菇	5 朵
金華火腿	20 克（切薄片）
生抽	1 茶匙
糖	1/2 茶匙
鹽	1 茶匙
薑汁、紹興酒	各 1 湯匙
胡椒粉	1/2 茶匙
生粉	1 湯匙
麻油	1/2 茶匙
葱白	2 條（切絲）

 烹調心得

蒸甲魚要蒸得滑，秘訣在於調味料要分次序先後加入，先用薑汁酒及調味料等入味，然後才加生粉鎖住肉汁，最後才放油和麻油加以潤滑。

1. 甲魚宰好沖洗一下，用攝氏 60 度的熱水燙過，擦去甲上的黑衣，起出甲魚肉斬成件，放在清水中浸 15 分鐘使之盡透血水，再洗淨瀝乾。甲魚殼洗淨備用。把甲魚肉用大沸水汆燙 1 分鐘，瀝乾。

2. 冬菇浸透，去蒂，斜切厚片。火腿用清水泡 5 分鐘，瀝乾。

3. 把生抽、糖、鹽、薑汁、紹興酒、胡椒粉等調味料拌勻，放入甲魚醃 15 分鐘。然後再加 1/2 湯匙生粉拌勻片刻，最後加麻油和 1 茶匙油再拌勻。

4. 把冬菇和火腿加入甲魚中稍作拌勻，把甲魚殼翻轉，放在大碗中，在殼內放入拌好的材料，隔水蒸 25 分鐘後移出，倒出汁水留用。

5. 用大碟覆蓋在甲魚殼上，倒扣在大碟中，用 1/2 湯匙生粉把汁水勾芡，倒在水魚上，撒上葱白絲，即成。

＊英文食譜見第 190 頁

品味上海菜 宴客菜

魚唇

　　很多人都不知道「魚唇」究竟是甚麼東西，往往認為魚唇是大魚的口唇，這是被誤導了。魚唇作為食材，已經有千多年了，最早的記載在唐代，名稱與現在不同；清代的《調鼎集》稱為「魚沖」，列作著名的「中八珍」食材之一；在《清稗類鈔》中更有「魚唇席」的記載，但近代的烹飪書籍，卻很少有涉及魚唇。

　　清代廣州成為對外經商口岸，粵菜中的魚翅和魚唇成為達官貴人們盤中的食材，其他外省菜系因對魚唇甚少接觸，百年下來也就淡忘了，惟有上海菜除外。上海開埠以來，經歷兩百多年來的經濟繁榮，孕育出海納百川的上海菜，上海人生活得精緻考究，當然不會忘記魚唇。

　　古代的魚唇，是用鯊魚面部的皮層加工而成，產量比魚翅還低，因為珍貴，平民百姓對它認識甚少。因為魚翅市場百年走俏，漁夫捕獲鯊魚之後，只懂得割下魚鰭，其他部分就拋回大海，所以魚唇的貨源幾近絕跡；而同時，古老的乾曬魚唇工藝漸漸失傳，對這種鯊魚面部皮層魚唇的認識，也就隨着時光流逝而失去了。

　　有一種民間俗稱「黃膠翅頭」，是生長在西沙群島及菲律賓一帶水域的鯊魚，千年以來它的魚翅都是宮庭貢品。它是上下兩端魚鰭之間的肉膜，全塊無骨無翅，形狀似一塊船帆的尾部，皮層特別厚而軟滑；據說皇帝吃後能提升活力體能，明朝皇帝賜名「魚唇」，從此鯊魚尾部的皮層，不論厚薄，都叫做魚唇。但是，由於亞洲水域的黃膠鯊魚（長鼻鋸齒鯊）已經絕跡，黃膠魚唇的貨源稀缺，現在還剩少量的是南美洲水域的黃膠魚唇。

魚唇頭

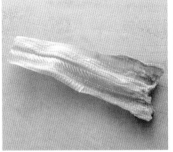

牙揀尾魚唇

老五羊魚唇

天九魚唇非常厚身

五羊魚唇

　　清朝以來的粵菜廚師，為了應付客人的需要，退而求其次，用群翅頭的中間皮層來充當魚唇，叫做「群翅皮」，浸發後晶瑩剔透，柔軟而有彈性，含豐富天然膠質，爽清的口感像花膠。兩代人過去後，新一代的廚師，只見過上一代的師傅浸發這些翅皮，就先入為主認為這就是魚唇。（部分資料來源參考《粵廚寶典》中的候鑊篇）

　　「群翅皮」現在已絕跡香港，現在西環海味市場上，用天九翅皮做的「天九魚唇」為最厚身，浸發起來像塊花膠，膠質豐富，當然價錢也最貴；還有用烏羊翅皮做的「魚唇頭」，雖然大部分為大酒家收購，但也是可以買到的。次一級的，還有一種「五羊魚唇」或「老五羊魚唇」，為上海菜和江浙菜餐館採用，製作魚唇的菜式。

品味上海菜 ❀ 宴客菜

香港的海味店出售一些寫着是「魚唇」的平宜貨，是來自小鯊魚的尾鰭皮層，是所謂「牙揀尾魚唇」。這種魚唇浸發後，比魚唇頭薄身，雜質也比較多，發好後要小心撕走。這種魚唇也含天然膠質，而價錢是魚唇頭的四分之一，一般家庭用來煲湯也是不錯的。

魚唇是相對價平而容易操作的海味食材。買回來的乾魚唇，樣子不討好，像塊乾木片，腥味較重；但只要懂得浸發，腥味很容易除去，可以烹調出矜貴的菜式，例如做魚唇羹、蟹粉燴魚唇、魚唇紅燒肉等等。

魚唇 的 浸 發 方 法

已浸發和乾的天九魚唇

把乾魚唇直接放蒸碟上，隔水蒸 30 分鐘，取出浸清水 10 至 12 小時已脹身，取出沖水，再蒸一次 30 分鐘，放入雪櫃用冰水再浸 12 小時，這時魚唇已完全膨脹及有彈性，可用作炆、燉、紅燒、煲湯。

已發起的魚唇頭

魚唇紅燒肉

STEWED PORK BELLY
WITH
SHARK SKIN

6-8 人份

準備時間 **24** 小時　烹調時間 **2** 小時

材料

五花腩	600 克
乾魚唇	100 克
薑片	50 克
白糖	30 克
上海豆瓣醬	1 湯匙
八角	2 粒
紹興酒	4 湯匙
紅燒醬油	2 湯匙
鹽	1 茶匙
冰糖	40 克

做法

1. 乾魚唇預先發好，切成大塊，備用。

2. 五花腩用冷水鍋汆水滾 3 分鐘，取出沖水，切 3 厘米方塊。

3. 用白糖炒糖色（見第 10 頁）。

4. 燒熱 2 湯匙油，爆香薑片和豆瓣醬，放入豬肉大火兜勻，加入八角和紹興酒，再加水至覆蓋豬肉，煮沸，轉中火，加蓋燜 30 分鐘。

5. 放入紅燒醬油、鹽、冰糖和煮好的糖色，煮沸，收慢火燜 45 分鐘。放入魚唇，再燜 15 分鐘，大火收汁即成。

＊英文食譜見第 191 頁

楓涇丁蹄

BRAISED PORK
KNUCKLE

楓涇，是上海金山縣的一個江南水鄉古鎮，卻因百年陳湯煮丁蹄而遠近馳名。相傳在清代道光年間，楓涇已是個聞名的集鎮，鎮上菜館林立，其中有丁姓兄弟倆，開了一家叫「丁義興」的飯店，但生意不好。咸豐二年（西元 1852 年），丁氏兄弟為了打開生意悶局，商量研究後，用太湖出產的著名楓涇豬的肘子（前腿連蹄），幾經改良，烹製出一道有獨特風味的滷煮肘子，結果大受食客歡迎，飯店的生意一炮而紅，門庭若市，不少人慕名而來品嚐，更被食客冠以「丁蹄」的名稱。

與一般的滷製食物不同，楓涇丁蹄所用的香料品種不多，只是用了丁香和桂皮，刻意地突出了這兩種香料的香味，食後為之齒頰留香，別具風韻，所以能在各省眾多滷製食品中脫穎而出。

楓涇丁蹄的誕生，距今已有 160 多年了，到了清宣統二年（1910 年），丁蹄榮獲南洋勸業會銀牌，並遠銷國外；1935 年再榮獲巴拿馬國際博覽會金獎，之後更獲獎無數。

品味上海菜 ✿ 宴客菜

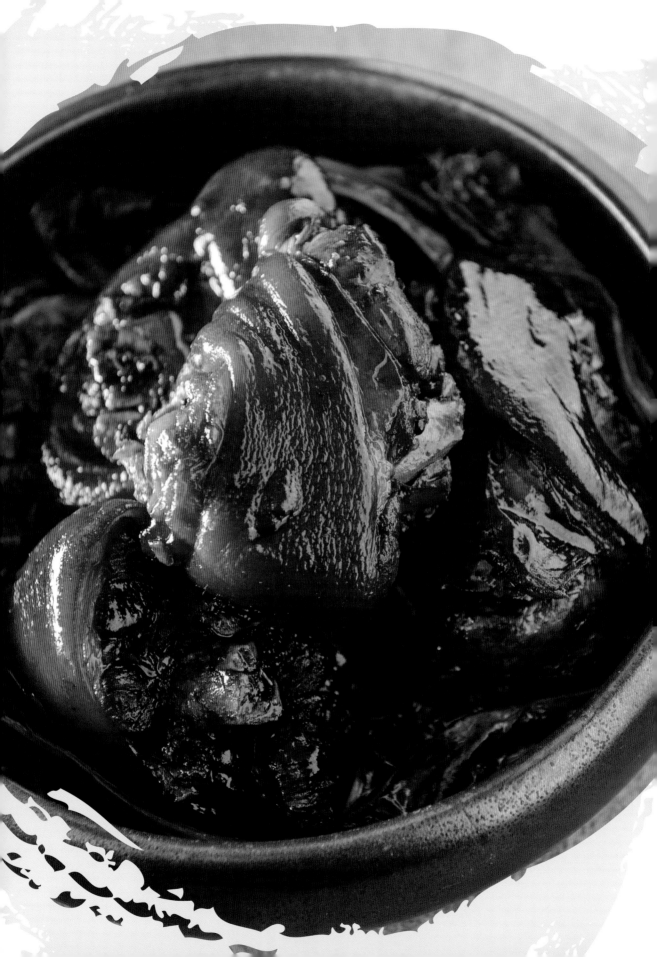

8 人份

準備時間 10 分鐘

烹調時間 1.5 小時

材料

肘子（豬前蹄上截）	2 隻
白糖	3 湯匙（炒糖色）
丁香	3 克
桂皮	6 克
薑片	30 克
紹興酒	125 毫升
紅燒醬油	3 湯匙
鹽	1 茶匙
冰糖	40 克

做法

1. 把肘子的皮毛刮洗乾淨，每隻橫刀砍為兩段成 4 塊，汆水，瀝乾。

2. 炒糖色備用（見第 10 頁）。

3. 丁香和桂皮放在香料袋內。

4. 在鍋裏燒熱 2 湯匙油，爆香薑片，放入香料袋、肘子，加紹興酒、醬油、鹽、冰糖、糖色和水至覆蓋材料，沸煮 3 分鐘，轉小火，加蓋燜至肘子軟糯，取出香料袋，大火收汁，即可裝盤。

＊英文食譜見第 192 頁

品味上海菜 🕸 宴客菜

紅燒河鰻

BRAISED EEL

　　河鰻即白鱔，上海菜中的紅燒河鰻，具有本幫菜濃油赤醬、味道偏甜偏濃的特色；要做到酥而不爛，入口即溶，加上顏色亮麗，令人垂涎三尺。

　　浙江寧波菜中，有一味傳統名菜叫做鍋燒鰻，做法與上海的紅燒河鰻相似，但醬油用得較少，而且加入了醋和桂皮，着重清鮮。福建菜中也有類似菜式，特色是重糖味，煮好之後盛起，待冷卻之後將中間的鰻魚骨抽去，然後蒸熱打芡上菜，稱為蔥燒通心鰻，是一道講究的宴客菜。

　　但是，這道紅燒河鰻，真的要做到酥而不爛，皮肉完整，並不容易。可惜不少餐館所做的，上桌時已是皮開肉綻，不忍卒睹。清康熙時期，著名文學家、美食家袁枚是浙江人，所著《隨園食單》中，記載「紅煨鰻」，用「煨」字代替「燒」字，即強調慢火煮至收汁的意思，文曰：「鰻魚用酒水煨爛，加甜醬代秋油，入鍋收湯煨乾，加茴香、大料起鍋。有三病宜戒者：一皮有皺紋，皮便不酥；一肉散碗中，箸夾不起；一早下鹽豉，入口不化。揚州朱分司家製之最精。大抵紅煨者，以乾為貴，使滷味收入鰻肉中。」最後那句最重要，意思是汁水要下得適量，醬汁逐漸收緊，吸收入鰻肉中，而濃汁緊緊包着鰻魚，這就是古人做紅煨鰻的秘訣之一。

另一個秘訣是豬油，在小火慢煮的過程中，要加入幾次豬油，豬油的黏性重，包住鱔皮，起了潤滑和保護作用，使鰻魚的皮不容易破裂。經過長時間的烹調，鰻魚皮上的脂肪會溶化在汁醬中，與豬油混為一體，使菜式色澤紅潤光亮，口感軟糯。

　　這是一道矜貴的請客菜式，看似很難做得好，其實只要細心和耐心，便人人都可以做到，絕對是一道甘腴可口的美食。

7-8 人份

準備時間 15 分鐘

烹調時間 50 分鐘

材料

河鰻 / 白鱔 ………… 1 條
（約 900 克）
鹽 ………………………… 1 湯匙
薑 ………………………… 50 克
蔥 ………………………… 4 條
豬油 ……………………… 4 湯匙
紹興酒、糖 ……… 各 2 湯匙
紅燒醬油 ……………… 2 湯匙

烹調心得

河鰻即白鱔，紅燒河鰻要做得又美觀又好吃，首先要選擇較幼身的河鰻，粗大的河鰻因為需要比較長時間煮，魚皮便容易破裂。

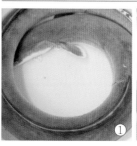
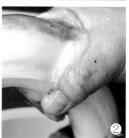

熱水浸泡白鱔　　　　清洗白鱔

做法

1. 白鱔宰後用 500 毫升沸水加 250 毫升冷水浸泡約 2 分鐘（圖 1），用手抹掉皮上的黏液（圖 2），再用鹽抓洗後用清水洗淨。
2. 切掉頭尾，只取中間大小比較均勻的部分，切成段長約 6 至 7 厘米。
3. 薑切片，蔥切段，備用。
4. 在鑊中下 2 湯匙豬油，先把薑片和蔥段爆香，放入河鰻和酒兜勻。
5. 加 500 毫升水，大火煮沸，轉小火，15 分鐘後放入糖和紅燒醬油，一邊煮一邊逐步加入少量豬油，煨煮約 30 分鐘至全熟酥爛。
6. 大火收汁時，再加一次豬油拌勻，使之更加油潤，即可盛碟。

＊英文食譜見第 192 頁

乾苔條

苔條粉

　　苔條是上海人和江浙人喜歡吃的食物，著名的菜式有苔條魚、苔條花生、苔條蝦仁、苔條明蝦、苔菜拖黃魚、苔菜豆腐，還有寧波人最喜歡吃的苔菜年糕和台州的海苔餅等。廣東人常吃紫菜、海帶和海草，當人們在江浙或上海館子吃到苔條的菜式，覺得很好吃，但不認識苔條，從近似紫菜的香味中，只知道是海的產物。

　　苔條是天然野生的海藻，又叫做苔菜，或條苔藻，在中國東海、黃海的海邊天然生長。苔條的藻體長約50厘米，有一條細長的主枝，上面有很多更幼的分枝，新鮮的苔條呈深綠色，曬乾之後是暗綠色。苔條營養豐富，含蛋白質、藻膠和多種維他命，苔條的蛋白質中富含各種人體需要的氨基酸。中醫認為苔條性寒，具清熱解毒的功效。用天然的乾苔條，自己動手製成苔條粉，無任何添加劑，是理想的綠色健康食物。

　　買回來的乾苔條，形態像一把亂頭髮（圖1），大大的一大束，散發着濃濃的海藻的味道。要注意的是，乾苔條不要用水洗，以保持海水的味道。用剪刀把乾苔條剪成數段，放在白鑊中，開非常小的火，不加水不加油，乾鑊慢慢焙走濕氣，同時用手捏碎乾苔條。當乾苔條全部捏碎，倒出攤開放至涼。然後用篩子把大粒的硬條及碎石蠔殼篩走，最後還要用手再次檢查，把所有手指感覺到的硬物揀走，直至苔條成粉狀。處理好的苔條粉（圖2），倒入玻璃瓶中，放雪櫃可保留半年，隨時可以用來做菜。

　　用苔條做菜，還要注意的兩點是，苔條的海水味很重，有天然的鹹味，做苔條的菜一般不用加鹽；苔條不宜耐煮，更容易燒焦，天然的味道便會消失。

　　苔條在南貨店可以買到。

苔條鳳尾蝦

FANTAILED PRAWNS
WITH
SEAWEED

 2-4 人份

準備時間
10 分鐘

烹調時間
3 分鐘

材料

大蝦	4 隻
薑汁	1 湯匙
葱	1 條（切段）
白胡椒粉	1/4 茶匙
鹽	1/4 茶匙
紹興酒	1 茶匙
生粉	2 茶匙
雞蛋白	1/2 個
麵包糠	3 湯匙
苔條粉	2 湯匙

＊注：苔條粉的做法，見第 127 頁。

做法

1. 把大蝦去掉蝦頭，剝去蝦殼（留蝦尾），洗淨。

2. 從蝦腹剖開至脊背（不切斷），挑出蝦腸，再把蝦肉張開壓成片狀，用薑汁、葱段、白胡椒粉、鹽和紹興酒醃 10 分鐘取出，用廚紙吸去水分，葱段不要。

3. 在蝦肉中間剖開一個口，把帶殼的蝦尾從蝦腹的開口中穿過去。

4. 先把蝦肉抹上一層生粉，再拌入雞蛋白，最後用麵包糠拌勻。

5. 在鍋內用大火把 750 毫升油燒至中溫（約 150℃），把每一隻蝦用手提蝦尾把蝦放入油鍋內，炸至轉色定型取出，大火升高油溫（約 170℃），把蝦放入油中再炸 5 秒， 取出瀝油，灑上苔條粉，即成。

＊英文食譜見第 193 頁

品味上海菜 ❀ 宴客菜

苔菜年糕

GLUTINOUS RICE CAKE
WITH SEAWEED

2-4 人份

準備時間 10 分鐘　烹調時間 5 分鐘

材料

寧波水磨年糕	300 克
苔菜	5 克
豬油	3 湯匙
白糖	1 湯匙

*注：苔條粉的做法，請參考第 127 頁。

材料選購：
寧波水磨年糕在南貨店有售。

做法

1. 年糕沖水洗淨，用刀斜切成半厘米厚片。

2. 苔菜用白鑊低溫烘乾，用手捏成粉狀。

3. 在煲裏燒熱 1 公升水至出現蝦眼水（約 90℃），熄火，放下年糕，泡約 2 分鐘至稍軟，瀝乾，再用 1 湯匙油把年糕拌勻，使每一塊都沾上油。

4. 在鑊裏燒熱 3 湯匙豬油，放入年糕，加白糖，用筷子把年糕分開，炒至稍為變黃，熄火，拌入苔菜粉即成。

＊英文食譜見第 193 頁

品味上海菜 宴客菜

砂鍋餛飩雞

WONTON
AND
CHICKEN CASSEROLE

這道湯菜味道鮮美。紹菜吸收了火腿、雞肉的精華，是最快被「掃清」的配菜。

4 人份

準備時間 30 分鐘　　烹調時間 30 分鐘

 材料

光雞	1/2 隻
金華火腿	50 克
絞豬肉	150 克
生抽	1 茶匙
鹽	1/2 茶匙
糖	1/2 茶匙
紹菜	300 克
生粉	1 湯匙
麻油	1/2 湯匙
上海餛飩皮	200 克
清雞湯	250 毫升

 做法

1. 半隻雞洗淨，放滾水汆水 1 分鐘，立即撈起，瀝乾。

2. 火腿切片放在小碟，加水過面，蒸 10 分鐘，把水倒去，火腿備用。

3. 把絞豬肉和生抽、鹽、糖及 2 湯匙水拌勻，醃 15 分鐘。

4. 紹菜焯水至軟，過冷河。把菜切碎後再擠出水分，加入絞豬肉。

5. 加入生粉，向一方向攪拌，加麻油，再拌勻成餛飩餡料。

6. 用上海餛飩皮把餛飩包好，燒開大鍋水，放入餛飩，水滾餛飩浮起時，加入半杯冷水，待水再滾時如果餛飩浮起，表示餛飩已經熟透，取出。

7. 在砂鍋裏煮滾清雞湯，放入雞和火腿，加水至完全覆蓋材料，沸煮 1 分鐘，加蓋，熄火，浸 15 分鐘。加入煮好的餛飩，再煮沸即成。可加鹽調味。

＊英文食譜見第 194 頁

品味上海菜 ❀ 宴客菜

荷葉八寶雞

STUFFED CHICKEN
WRAPPED IN
LOTUS LEAF

八寶鴨這道菜，據說是源自蘇州的傳統名菜糯米鴨子，據乾隆三十年南巡時的《江南節次照常膳底檔》中記載：「正月二十五日，蘇州製造普福進糯米鴨子」。清代的《調鼎集》也記錄了八寶鴨的製作方法，乾隆時期，八寶鴨已是清代宮廷名菜。

一隻鴨再加糯米和八種餡料，重達斤餘，估計八人以上才可以吃得完，於是民間就改為做八寶雞，這一改就是百多年了。做八寶雞這道菜而著名的，是在上海老城隍廟附近的老字號老榮順菜館，上世紀40年代初，老榮順搬遷至麗水路，改名為上海老飯店，成為上海地標性的傳統飯店。

所謂八寶餡料，是比較隨意的，但一定是由八種材料組成，糯米飯、冬菇、蓮子和栗子這四種是一定要有的，再隨意配合其他食材如江珧柱、蝦米、肉丁、火腿丁、筍丁、雞胗丁、青豆等等。

無論是八寶雞還是八寶鴨，都是把雞鴨褪骨，把餡料釀在皮囊中，之後入爐乾蒸，然後煮汁澆上去。由於蒸熟之後，要把這好幾斤重的無骨而裝滿餡料的雞鴨，拿出倒扣在盤碟上，還要把汁水先留出來，這個廚房功夫實在不容易，必須是受過訓練的廚師，手力不夠的家庭主婦，一不小心就會把雞擲爛了。有見及此，我們家便改做荷葉八寶雞，有荷葉的承托，由蒸鍋中取出就容易多了。

6-8 人份

 準備時間
1.5 小時

烹調時間
1.5 小時

材料

糯米	160 克
金華火腿丁	50 克
栗子肉	12 個
光雞	1 隻
鹽	1.5 湯匙
薑汁	1 湯匙
紹興酒	1 湯匙
乾荷葉	2 張
冬菇	2-3 朵
江珧柱	30 克
蝦米	20 克
蓮子	30 克（去芯）
花生	20 克
老抽	1 湯匙

趁熱用筷子挑鬆糯米飯

將糯米飯釀入雞腔內

用荷葉把雞包裹

做法 . . . .

1. 糯米用熱水加 1/2 湯匙鹽泡 1 小時，沖淨，
   瀝乾。

2. 金華火腿丁用水洗淨，蒸 10 分鐘。

3. 栗子肉去衣洗淨，蒸 1 小時。

4. 光雞洗淨，去骨，用 2 茶匙鹽、1 湯匙薑汁
   和 1 湯匙紹興酒醃 30 分鐘。

5. 乾荷葉洗淨浸泡至軟身。

6. 冬菇泡軟，去蒂，切粒。

7. 江珧柱用水泡軟，撕成絲；泡江珧柱水留用。

8. 蝦米用清水泡軟，用時瀝乾。

9. 蓮子和花生洗淨，蒸 15 分鐘後取出。

10. 用 2 湯匙油把火腿、冬菇、江珧柱、蝦米、
    花生、栗子和蓮子炒約 1 分鐘，再放入糯
    米、1 湯匙老抽、1 茶匙鹽和泡江珧柱水拌
    勻，放在大碟內蒸約 45 分鐘或至糯米全熟
    後放涼。

11. 把糯米飯釀入雞腔內，再用荷葉把雞包住，
    大火蒸 30 分鐘即成。

＊英文食譜見第 195 頁

翡翠素方

VEGETABLE ROLL

　　請客菜式中，素菜是不能或缺的部分，無論客人信奉那一種宗教，都不會抗拒進食素菜；現代人講求健康，素菜總是受歡迎的。一道白焯或清炒時菜，簡單快捷，卻未能表達心思，不如多花一點工夫，做一個精巧的「翡翠素方」，來贏取客人的讚賞吧！

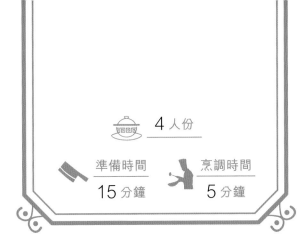

4 人份

準備時間
15 分鐘

烹調時間
5 分鐘

材料

腐皮	1 張
菠菜	450 克
冬菇	4 朵
牛肝菌粉	1 湯匙
雞蛋	1 個
芝麻	1 湯匙
鹽	3/4 茶匙
麻油	1/2 茶匙

品味上海菜 ✿ 宴客菜

烹調心得

1. 如有患尿酸高的人，不適宜吃菠菜，可改用莧菜，或其他綠色的菜，菜中也可以加入紅蘿蔔絲。

2. 牛肝菌粉可以增加蔬菜的鮮味，是做素菜的好幫手，但也可以不加入菌粉。

做法 · · · · · · · · · · · · · · · ·

1. 菠菜洗淨，用水灼軟立刻撈出，
   沖冷開水再擠乾，切去約 4 厘米
   菜梗，餘下的菜葉和嫩的菜梗略為
   切碎。

2. 冬菇浸透，去蒂切成幼絲，瀝乾水
   分；雞蛋打勻成蛋漿；芝麻用白鑊
   炒香。

3. 把菠菜和冬菇絲、牛肝菌粉、芝麻
   加鹽和麻油拌勻，平均分為四份。

4. 把圓形的腐皮對摺兩次，用刀把圓
   的邊切除（圖 1），裁成四方形，
   再切為四張四方形腐皮。

5. 把一張腐皮抹上一層蛋漿（圖 2），
   鋪上一份餡料（圖 3），再捲成扁
   平的長方形包（圖 4）。同樣處理
   其他的腐皮。

6. 用平底鑊開小火，加 1 湯匙油，把
   素方放下，慢慢地煎至兩面金黃，
   取出用紙瀝油，切開上碟，即成。

＊英文食譜見第 196 頁

湯羹

水為鄉，蓬作舍，魚羹稻飯常餐也。

——唐·李珣《漁歌子·楚山青》

上海菜的湯與羹，
雖然沒有廣東湯品般繁多，
但勝在性格鮮明，
而且上得廳堂、入得弄堂，
讓人品嚐過後難以忘懷。
這篇章介紹的湯羹，有湯有料，
一碗酸辣湯加上熱氣騰騰的包子，
就是平民百姓暖胃飽肚的早午餐。

油豆腐粉絲湯

TOFU PUFF AND MUNG
BEAN VERMICELLI SOUP

〈油豆腐〉市中白水常鹹醉，寺裏清油不碑禪。

——清・李調元《豆腐詩》

　　油豆腐粉絲湯，上海人叫做油豆腐線粉湯，這是由家常菜發展出來的大眾化小吃。上世紀二三十年代，很多外地人到上海做勞力，他們一日三餐都在弄堂邊上的飯攤解決，一碗熱氣騰騰的油豆腐粉絲湯，用料簡單，但有湯有料，早餐吃一碗，精神力氣就來了；下午或宵夜吃一碗，不多不少，正好安慰一下軟弱的食慾。

　　油豆腐粉絲湯的用料，基本上就是油豆腐（豆腐泡）、粉絲和湯，講究的做法，會為它特意煲個雞湯或排骨湯，還加上些有肉餡的百葉包，用幾片榨菜吊出鹹鮮的滋味。

 4 人份

準備時間
5 分鐘

烹調時間
1 小時

材料

排骨	150 克
油豆腐（豆腐泡）	8 個
粉絲	20 克
榨菜	30 克
鹽	1 茶匙
胡椒粉	1/8 茶匙

做法

1. 排骨汆水,過冷水,加 1.5 公升水煲 30 分鐘成排骨湯。

2. 把油豆腐切開成兩半,用淡鹽水沸煮 5 分鐘,瀝乾。

3. 粉絲用水浸泡 5 分鐘,撈出。

4. 榨菜洗淨,切片。

5. 把排骨湯煮沸,下油豆腐和榨菜煮 5 分鐘,放入粉絲再煮 3 分鐘,加鹽和胡椒粉調味,即成。

＊英文食譜見第 196 頁

品味上海菜 ✿ 湯羹

海參蝦仁豆腐羹

SEA CUCUMBER,
SHRIMPS AND TOFU SOUP

莫將腐乳等閒嘗，一片冰
心六月涼。不日堅乎惟日白，
勝他什錦佑羹湯。

——清・林蘭痴《咏豆腐》

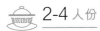 2-4 人份

 準備時間 10 分鐘　　烹調時間 5 分鐘

材料

水發海參	250 克
嫩豆腐	250 克
蝦仁	100 克
清雞湯	250 毫升
薑汁	1 湯匙
白胡椒粉	1/8 茶匙
鹽	1/2 茶匙
馬蹄粉	2 湯匙
雞蛋白	1 個

 做法

1. 先把水發海參切成 1 厘米大小，汆水，瀝乾。
2. 蝦仁洗淨後，切成 1 厘米大小。
3. 嫩豆腐也切成 1 厘米大小。
4. 在鍋裏煮沸清雞湯和薑汁，放入海參粒、豆腐粒和蝦仁，同煮約 3 分鐘，再加白胡椒粉和適當的鹽調味，最後用馬蹄粉加水勾芡，熄火。
5. 在羹的上面，把打勻的雞蛋白隔着漏勺，把蛋白液流到羹上，等 1 分鐘後，輕輕拌勻即成。

＊英文食譜見第 197 頁

145

品味上海菜 ❀ 湯羹

薺菜肉丸湯

SHEPHERD'S PURSE AND
MEATBALL SOUP

去過上海的人，很多都會吃過薺菜肉餛飩，每年的冬季和春季，上海大小飯店都會爭相推出薺菜的季節菜式，薺菜清香可口，非常受歡迎。

《詩經》中說薺菜「誰謂荼苦，其甘如薺」，中醫認為，薺菜有清熱、利尿、止血等功效。薺菜原來是江南地方的一種野菜，人們吃了三千多年，變成了種植的蔬菜，本來只有在春季當造的薺菜，因為有了溫室種植，現在幾乎一年四季都可以買到薺菜。

4 人份

準備時間 15 分鐘　烹調時間 10 分鐘

材料

薺菜	600 克
絞豬肉	200 克
生抽	1 茶匙
糖	1 茶匙
生粉	1 茶匙
雞蛋	2 個
雞湯	500 毫升
豬油	1 茶匙
鹽	1/2 茶匙

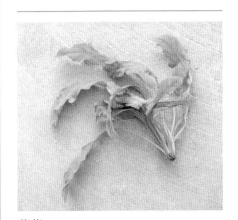

薺菜

做法

1. 薺菜洗淨，切掉菜根。

2. 大火煮一鍋沸水，放入薺菜焯熟，撈起用清水沖過，瀝乾水分。

3. 鍋中加 1 湯匙油，中火燒熱，放入薺菜略為兜炒，和雞湯一同放進攪拌機攪拌成薺菜湯，再倒回鍋裏。

4. 把豬肉、生抽、糖和生粉拌勻，加 2 湯匙水，再加 1/2 個雞蛋白，朝一個方向攪拌至起膠。用手沾水把混合好的肉做成肉丸。

5. 煮沸薺菜湯，放下肉丸煮 2 分鐘，加入豬油和鹽，熄火；把餘下的雞蛋打成漿，均勻倒下湯中，即成。

＊英文食譜見第 197 頁

品味上海菜 ❀ 湯羹

酸辣湯

| HOT AND SOUR SOUP

　　酸辣湯在香港流行了幾十年，一點也不覺得它「老土」，很多年輕人都喜愛它。我們在小孩子年代，跟媽媽去上海館子吃飯，就認識了酸辣湯，一直都知道它是上海菜；因為在香港的上海館子都一定有酸辣湯，就像吃上海菜肉雲吞那樣的理所當然。

　　原來，酸辣湯源自四川，不過可能您搜遍四川菜食譜，麻婆豆腐、螞蟻上樹、大千雞、回鍋肉、水煮牛肉、宮保雞丁、酸菜魚……就是沒有酸辣湯，我也曾百思不得其解，問一位四川朋友，他笑着說酸辣湯哪裏算得上是「菜」？它只是四川的街頭小食，是平民老百姓的食物。在四川，大街上賣包子、饅頭、大餅的小食檔，有些就同時賣酸辣湯，一大碗熱氣騰騰的酸辣湯，加個大饅頭，就是一頓飽肚的午餐。四川人相信酸辣湯能行血祛濕，在四川這種濕熱的氣候，不失為價廉物美又健康的食品。

　　酸辣湯其實不應該叫做湯，應該稱為「羹」，因為它是連湯料一起吃，而且是用埋芡變稠的，不過人們都已習慣叫它做酸辣湯。酸辣湯的酸味來自醋，但辣味並非來自辣椒，而是來自白胡椒粉，如果愛吃更辣，就自己加紅辣椒油吧！

4-6 人份

準備時間 20 分鐘　烹調時間 10 分鐘

 材料

熟豬肚	100 克
豬紅	200 克
冬菇	4 朵
乾木耳	5 克
硬豆腐	1/2 塊（125 克）
筍	1 隻
雞蛋	1 個（打勻）
鹽、糖	各 1/2 湯匙
馬蹄粉	2 湯匙
鎮江香醋	5 湯匙
胡椒粉	1 湯匙
麻油	1 茶匙

 烹調心得

1. 醋用鎮江香醋，也可以用白醋。
2. 不吃內臟的，可取消熟豬肚。
3. 新鮮筍可以用罐頭筍代替，汆水是要把筍的苦味去掉。

做法

1. 熟豬肚切絲。豬紅切絲，汆水。
2. 冬菇、木耳泡軟後切絲。
3. 豆腐加 1/4 茶匙鹽，用熱水泡 15 分鐘，瀝乾，切絲。
4. 筍剝去外殼，用刨刨去表皮，切成 4 厘米段，把每一段切開兩邊，再斜切成薄片，再切成細絲。用水加半茶匙鹽把筍絲汆水 2 分鐘，瀝乾。
5. 在煲中煮沸 1 公升水，放入肚絲、冬菇絲、木耳絲、筍絲和豬紅，加入糖和 1 茶匙鹽，煮沸，用筷子輕輕推散（不要攪拌）。
6. 用少許水把馬蹄粉溶化，徐徐倒入湯內，邊倒邊攪拌。
7. 放入豆腐煮滾，轉小火，把雞蛋液經過網篩，慢慢打圈子倒進湯內，不要攪拌，便成蛋絲，熄火。
8. 放入醋和胡椒粉，攪拌，加入麻油即成。

＊英文食譜見第 198 頁

飯麵

小時候，
母親帶着我和姐姐上館子，
經常點嫩雞煨麵、上海粗炒或蔥油拌麵，
是我美好的美食回憶。
上海菜飯也是我家很喜愛吃的家常飯，
美食家父親堅持做菜飯一定要有些豬油，
那香噴噴的菜飯，
是難以抵擋的誘惑。

湯餅一杯銀絲亂，牽絲如縷王箸惜。

——北宋 • 黃庭堅 《過山寨詩》

葱油拌麵

NOODLES WITH
SPRING ONION OIL

這是一道做法簡易又美味的麵食。葱油可預先做好一瓶放在雪櫃，需要用時將它燒滾淋在麵上拌勻即可。

1 人份

準備時間
10 分鐘

烹調時間
10 分鐘

材料

上海幼麵	100 克
清雞湯	2 湯匙
生抽	1 茶匙
老抽	1 茶匙
炸葱	隨量
葱油	3 湯匙

* 葱油和炸葱的製法,見第 10 頁。

做法

1. 上海麵用水焯至熟,瀝乾水分,裝在大碗中。
2. 把雞湯、生抽、老抽拌勻成汁,淋在麵上,放上炸過的葱。
3. 燒熱葱油,淋在麵上,吃時拌勻。

＊英文食譜見第 199 頁

粢飯

SHANGHAI GLUTINOUS
RICE ROLL

在早上享用一個熱騰騰、餡料豐富的粢飯，為當天展開一個美好的序幕。

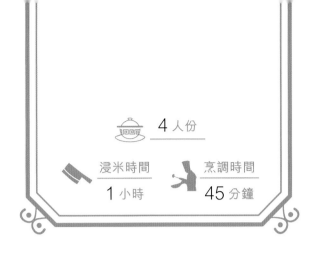

4 人份

浸米時間
1 小時

烹調時間
45 分鐘

材料

糯米	600 克
榨菜	150 克
炸油條	1 孖
豬肉鬆	1/2 杯
糖	1 茶匙

做法

1. 糯米用熱水浸約 1 小時，瀝水，放在蒸碟中撥至平均，隔水蒸約 45 分鐘至米熟，取出用木杓弄鬆飯粒，再分成 4 份。

2. 榨菜稍浸 5 分鐘，切碎，用 1 茶匙油稍為炒香取出，加糖拌勻，備用。

3. 將一孖油條撕成兩條，每條再剪成兩段，共 4 段。

4. 左手載上手套，把一塊乾淨的白布鋪在手掌上，再鋪上保鮮紙，抹上少許油，放入 1 份糯米飯，用杓稍為壓平，放入一段油條，加適量榨菜和肉鬆，把保鮮紙及白布捲起，捏成粗條型，放開白布，即成。

＊英文食譜見第 199 頁

品味上海菜 ✿ 飯麵

葱開煨麵

「葱開」是指葱油及開陽（蝦米），這是經典的上海家常湯麵。用蝦米、葱油、雞湯熬煮的麵湯，味道鮮甜，滋味無窮。

BRAISED NOODLES WITH
DRIED SHRIMPS AND
SPRING ONION OIL

2 人份

準備時間
5 分鐘

烹調時間
10 分鐘

材料

上海麵	300 克
蝦米（開陽）	50 克
葱油	2 湯匙
紹興酒	1 湯匙
淡雞湯	250 毫升
鹽	適量

* 葱油做法見第 10 頁。

做法

1. 蝦米洗淨，泡約 5 分鐘，瀝乾。
2. 上海麵用水焯一下，瀝乾。
3. 燒熱葱油，爆香蝦米，下酒、雞湯和 250 毫升水，煮沸，放入麵條，慢火煨 5、6 分鐘，用鹽調味即成。

＊英文食譜見第 200 頁

品味上海菜 ✿ 飯麵

上海粗炒

STIR-FRY SHANGHAI
THICK NOODLES

　　我母親是浙江紹興人，小時候，我和姐姐常常跟着父母上江浙館子吃午飯，反而比較少上茶樓吃點心。記得在五、六十年代，凡是有江浙口音的人，都被叫做上海佬、上海婆，而當年香港有不少上海館子，一律叫做上海舖。

　　很懷念那個年代的上海舖，店舖的明檔擺滿了五香燻魚、油爆蝦、醉雞、鳳尾魚、鹽水毛豆、醬蹄、西芹豆乾、五香花生，林林總總任君選擇。媽媽和我們幾乎必吃的是油豆腐粉絲湯、嫩雞煨麵，還有上海粗炒。

　　香港人喜歡吃的上海粗炒，在上海就叫做炒麵。弄堂裏的街坊小店，炒麵和菜肉餛飩都是平民小吃，普及程度等於香港的豉油皇炒麵。上海的炒麵只用粗麵，不用幼麵，所以不會刻意地叫做炒粗麵。香港人把它叫做上海粗炒，而不是炒粗麵，那是約定俗成的本地文化，來源不詳。外地人去到上海，通常是住酒店，出入上海的菜館，怪不得很少見到上海的炒麵。

　　其實做得好的上海粗炒，是半炆半炒的，不能像豉油皇炒麵那樣炒，否則那全麥的粗麵，就會不入味。

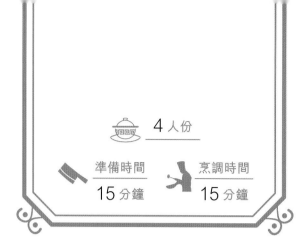

4 人份

準備時間
15 分鐘

烹調時間
15 分鐘

材料

上海粗麵（乾）	300 克
冬菇	3 朵
椰菜	300 克
瘦肉	150 克
鹽	1/2 茶匙
糖	1.5 茶匙
生粉	1/2 茶匙
生抽	1/2 湯匙
老抽	1.5 湯匙
麻油	1 茶匙

 ## 做法

1. 先把上海粗麵用水煮約 12 分鐘或至腍，再用清水洗淨麵上的澱粉，瀝乾。
2. 冬菇泡軟，去蒂，切絲；椰菜切細，洗淨。
3. 瘦肉切絲，加鹽、糖、生粉各 1/2 茶匙和 1 茶匙油拌勻。
4. 在小碗中放入老抽、生抽、1 茶匙糖和 200 毫升水或淡雞湯調成調味汁。
5. 在鑊裏燒熱 2 湯匙油，放入豬肉、椰菜、冬菇和麵，再把調味汁沿鑊邊淋下，煮沸後加蓋慢火焗 8 至 10 分鐘，掀蓋兜勻至完全收汁，再拌入麻油即成。

＊英文食譜見第 200 頁

上海菜飯

SHANGHAI VEGETABLE RICE

上海菜飯，是香港人的叫法；上海叫做豬油菜飯，加幾片鹹肉是豪華版。由清朝末年開始，上海成為當時中國重要的對外通商口岸之一，碼頭和倉庫的貨運非常忙碌，附近農村有很多窮人跑到這些碼頭和貨倉當苦力，這可能是中國第一代的入城民工。這些民工在上海賺取血汗錢，生活十分節儉，碼頭附近有些小販，就用最廉價的上海青（小棠菜）加在有鹽的飯中同煮，賣給這些碼頭工人吃，成為當時稱為「苦力飯」的第一代菜飯，因為加入豬油，後來就叫做豬油菜飯，登堂入室。1949 年中華人民共和國成立之前，上海經營豬油菜飯的小店和攤販，不下百家，最著名的就是創於 1921 年的美味齋，迄今已有九十多年歷史了。

上海菜飯是我家很喜愛吃的家常飯，美食家父親堅持做菜飯一定要有些豬油，那是菜飯香噴噴的原因，否則小棠菜缺油味寡，飯味就不是那一回事了。肥肉炸油後的豬油渣非常可口，混在菜飯中偶然脆口一下，人間美食也。很多人每天吃牛油麵包，其實牛油比豬油所含的膽固醇高得多，為何單要害怕偶然吃一次豬油呢？

4 人份

準備時間 | 烹調時間
20 分鐘 | 煮飯時間

 材料

白米	320 克
上海鹹肉	100 克
小棠菜	100 克
肥豬肉	30 克
蒜蓉	1 茶匙
鹽	1/2 茶匙

材料選購：
上海鹹肉可以在南貨店購買。

做法

1. 白米洗乾淨後瀝乾。

2. 上海鹹肉用清水泡浸 1 小時，再蒸 10 分鐘，然後切成 1/2 厘米厚片。

3. 小棠菜洗淨，在開水裏迅速汆一下，取出瀝乾，切碎。

4. 肥豬肉切粒。

5. 在鑊中，放進肥豬肉粒，加 2 湯匙清水，慢火煎至收水，開始煎出豬油，豬油渣拿出留用。

6. 把蒜蓉放入有豬油的鑊中略炒到出味，加入白米同炒約半分鐘。

7. 把炒過的米放在電飯鍋裏，放進鹽，加正常煮飯水量煮飯。

8. 當米飯開始收水時，放入鹹肉，繼續煮飯。

9. 最後把切碎的小棠菜和煮好的飯拌勻，加蓋焗 3 分鐘，撒上豬油渣，即成。

＊英文食譜見第 201 頁

# CONTENTS

豬油 | MAKING OF LARD

Refer to p.9

INGREDIENTS
300 g pork fatback, 2 cloves garlic, 30 g ginger slices

METHOD
1. Dice pork fatback and add to the wok together with 125 ml water, bring to a boil, reduce to low heat and allow pork fatback to fry slowly until brown. Remove pork, add garlic, ginger and fry until brown. Remove garlic and ginger. Transfer oil (lard) to a container and refrigerate after cooled.

2. The lard will turn a white color when cold.

葱油 | MAKING OF SPRING ONION OIL
Refer to p.10

INGREDIENTS
300 g spring onions, 500 ml oil

METHOD
1. Rinse spring onions and cut into 3 sections.

2. Heat oil to medium heat, put in stem sections and deep fry until stems turn slightly brown. Remove stem sections and put in the remainder of the spring onions. Slow fry until crispy and remove.

3. Spring onion oil should be refrigerated after cooled.

糖色 | MAKING OF CARAMEL
Refer to p.10

INGREDIENTS
2 tbsp sugar, 3 tbsp water

METHOD
1. Add sugar and 1 tbsp water into a wok and dissolve sugar over a medium low heat. Stir frequently. The dissolved sugar will first become large bubbles, then change gradually to small bubbles and syrupy. Reduce to low heat and keep stirring until syrup turns a deep brown color. Turn off heat.

2. Allow syrup to cool for about 1 minute and then put in the remaining 2 tbsp water. Mix well.

Note: Syrup is extremely hot and be very careful when making it.

香糟汁 | MAKING OF FRAGRANT DISTILLER'S GRAIN SAUCE
Refer to p.11

INGREDIENTS

150 g fragrant distiller's grain, 30 g ginger slices, 3 stalks spring onion (sectioned), salt, sugar

METHOD

1. Bring to a boil 1500 ml of water together with distiller's grain, ginger and spring onions, reduce to low heat and cook for 30 minutes. Cover and set aside for 8 hours.

2. Filter 2 to 3 times using a cotton cloth or bag to obtain about 750 ml of fragrant distiller's grain sauce. Flavor with 2 tbsp salt and 3 tbsp sugar. Refrigerate until use.

3. The sauce can be re-use once or twice, but should be re-boiled and add salt and sugar (2:3). Refrigerate after cooled.

烤四 麩喜 | SAUTÉED WHEAT KAOFU WITH VEGETABLES
Refer to p.16-19

INGREDIENTS

200 g wheat kaofu
podded soy beans
5 g dried black fungus
6 small dried black mushrooms
20 g dried day lily
50 g bamboo shoots
2 tsp ginger juice
10 g ginger, shredded
1 tbsp oyster sauce
1 tbsp Shaoxing wine
1 tsp sugar
1/2 tsp salt
1 tsp sesame oil

METHOD

1. Tear kaofu into small pieces. Bring to a boil 500 ml water with 1 tsp of ginger juice, add kaofu and blanch for 1 minute. Remove and flush under running water until cool, squeeze all water from the kaofu. Repeat process once.

2. Boil soy beans for 5 to 6 minutes.

3. Pre-soak dry black fungus, and cook in boiling water for about 2 minutes. Drain.

4. Soak mushrooms until soft, drain and remove stems.

5. Soak dry day lily for 10 minutes, rinse and drain.

6. Slice bamboo shoots and blanch for 2 minutes. Drain.

7. Heat 250 ml oil to medium heat (about 160°C), and deep fry kaofu until golden.Drain and press with a spatula to remove oil as much as possible.

8. Heat 3 tbsp of oil and stir-fry ginger and mushrooms over high heat, add kaofu, soy beans, bamboo shoot, day lily and black fungus, drizzle wine along the inside of the wok, add oyster sauce, salt and sugar, and toss. Stir in sesame oil before serving.

拌馬香蘭乾頭 | KALIMERIS AND TOFU SALAD
Refer to p.20-23

INGREDIENTS
300 g kalimeris
2 pcs dried tofu
1 tsp sesame oil
1/2 tsp salt
a few gouqi (for garnish)

METHOD
1. Rinse kalimeris. In a pot of boiling water, add 1/2 tsp oil and quickly blanch kalimeris, remove and rinse with cold drinking water. Squeeze excess water from kalimeris.
2. Rinse and blanch dried tofu.
3. Use kitchen towels to remove excess water from dried tofu and kalimeris, then chop and mix with salt and sesame oil.
4. Fill a round stainless steel mould with chopped tofu and kalimeris, press firmly, and push out onto a plate before serving.

TIPS
1. Blanch kalimeris only very briefly to maintain the bright green color.
2. Dried tofu is already cooked and only require a quick blanching.
3. Adding oil to the water to blanch kalimeris helps to maintain the color.
4. Rinsing kalimeris with cold drinking water keeps it crispy and helps to remove the grassy taste.
5. A 7 x 5.5 cm mold is the right size for this recipe.

瓣雪酥菜豆 | BROAD BEAN PASTE WITH PRESERVED POTHERB MUSTARD
Refer to p.24-26

INGREDIENTS
500 g frozen broad beans
100 g preserved potherb mustard, stem only
3 tbsp spring onion oil (see P.165)
250 ml unsalted chicken broth
2 tsp sugar
sesame oil

METHOD
1. Blanch broad beans, rinse and drain.
2. Rinse preserved potherb mustard, squeeze to remove water, and cut into small bits.
3. Heat spring onion oil, stir-fry broad beans, add chicken broth and cook until broad beans softened. Remove about 1/5 for later, and continue to stir-fry and press with the spatula to mash the broad beans.
4. Add preserved potherb mustard, reserved broad beans and sugar, and stir-fry until all the water has evaporated. Set aside to cool.
5. Smear the inside of a bowl or box with sesame oil, put in broad bean paste, press to smooth out surface, and refrigerate overnight. Turn over bowl and transfer contents to plate.

燻魚 | SPICED SMOKE FISH

Refer to p.27-29

INGREDIENTS

grass carp without fish head (about 500 g)
2 stalks spring onion, stems sectioned
30 g grated ginger
1/2 tsp salt
2 tbsp Shaoxing wine
2 tbsp Shanghai soy sauce
1/2 tsp five spice powder
2 tbsp Zhenjiang vinegar
1.5 pc pressed sugar, crushed
2 tbsp sesame oil

METHOD

1. Clean and rinse fish, drain, cut into thick slices and pat dry with kitchen towels.
2. Smash spring onion stems and put in a large bowl together with ginger, salt and 1 tbsp wine. Coat fish completely with sauce and marinate for 15 minutes.
3. In another small pot, put in 1 tbsp of wine, soy sauce, five spice powder, vinegar, pressed sugar and 4 tbsp of water, and add the marinating sauce and spring onion greens. Heat the sauce over low heat until the pressed sugar has melted, then pour into a large bowl.
4. Pat dry fish with kitchen towels and deep fry in 1 litre of oil until light brown. Remove and drain excess oil from the fish.
5. Dip fish immediately in the bowl of sauce one at a time and coat each side completely with the sauce, then place fish on a plate and brush with sesame oil.

糟毛豆 | FRESH SOY BEANS IN FRAGRANT DISTILLED GRAIN SAUCE

INGREDIENTS

300 g fresh soy beans
1 tbsp Sichuan pepper
1 tsp salt
1 tsp 5 spice powder
fragrant distilled grain sauce (see P.166)

METHOD

1. Clean soy beans and cut off the two ends of the pod.
2. Bring to a boil a pot of water, add Sichuan pepper, salt and 5 spice powder, and boil for 5 minutes. Put in soy beans and cook over medium heat for 20 minutes. Turn off heat. Remove soy beans after 10 minutes.
3. Pat dry soy beans with kitchen towels, soak in fragrant distilled grain sauce, refrigerate for 4 to 5 hours. Turn over soy beans twice.

糟
豬
肚

PIG'S STOMACH IN FRAGRANT DISTILLED GRAIN SAUCE

INGREDIENTS

1 pig's stomach,
2 tbsp corn starch
1 tbsp salt
2 tbsp white vinegar
2 tbsp crushed white peppercorn
fragrant distilled grain sauce (see P.166)

METHOD

1. Turn the fresh pig's stomach inside out, rub with corn starch, rinse, rub with salt and rinse again. Trim the fat off the inside of the stomach. Turn the pig's stomach inside out again. Rinse and cut into several large pieces.

2. Heat a large pot of water, put in pig's stomach, vinegar and crushed white peppercorn, and boil over high heat for 15 minutes. Reduce to medium low heat and cook for 1.5 hours or until soft. Rinse, drain and cut into strips.

3. Soak pig's stomach in fragrant distilled grain sauce for 24 hours.

糟
蝦

PRAWNS IN FRAGRANT DISTILLED GRAIN SAUCE

INGREDIENTS

300 g prawns
fragrant distilled grain sauce (see P.166)

METHOD

1. Rinse shrimps, blanch until done, drain and set aside to cool.

2. Trim shrimps and soak in fragrant distilled grain sauce for 4 to 5 hours.

酒糟鴨舌 | DUCK TONGUE WITH PICKLED SAUCE

Refer to p.35-37

INGREDIENTS
300 g duck tongue
30 Sichuan pepper
2 star anise
4 slices ginger
1 tsp sugar
1/2 tsp salt
3 stalks spring onion, sectioned
15 ml Shaoxing wine
3 tbsp pickled sauce
1/2 tsp sesame oil

METHOD
1. Wash and clean duck tongue, boil for 20 minutes and rinse with cold water. Drain.
2. Put Sichuan pepper and star anise in a spice pouch.
3. Boil ginger slices and spice pouch in 500 ml of water for 20 minutes. Discard ginger and spice pouch.
4. Add sugar, salt and spring onion, bring to a boil, then turn off the heat. When the sauce cools, add wine and pickled sauce.
5. Put duck tongue in pickled sauce and coddle for 2 to 3 hours before taking them out. Brush with sesame oil before serving.

糟雞 | CHICKEN MARINATED IN FRAGRANT DISTILLED GRAIN SAUCE

Refer to p.38-39

INGREDIENTS
1 chicken about 1200 g
300 ml fragrant distilled grain sauce (see P.166)

METHOD
1. Clean and rinse chicken. Drain.
2. Bring to a boil a large pot of water, add chicken with the breast facing up, and re-boil. Turn off the heat, cover and steep chicken in hot water for about 17 minutes. Turn the chicken with the breast facing down and steep for 5 minutes. Remove chicken to cool. Cut chicken into two halves.
3. Put chicken and the fragrant distilled grain sauce in a large food storage bag. Seal bag after squeezing out all the air. Refrigerate for 24 hours.
4. Remove chicken from the bag and cut into pieces. Drizzle fragrant distilled grain sauce on the chicken before serving.

蜜汁鳳尾魚 | DEEP FRIED GRENADIER ANCHOVY

Refer to p.40-41

INGREDIENTS

600 g grenadier anchovy
2 cloves garlic
10 g ginger
1/2 tsp salt
1/4 tsp ground white pepper
3 tbsp sugar
1 tsp fish sauce
1 tbsp Shaoxing wine

METHOD

1. Peel and chop garlic, and grate ginger.

2. Remove fish head and stomach, and clean and rinse fish.

3. Marinate fish with salt and ground white pepper for 20 minutes, pat dry with kitchen towel.

4. Heat 500 ml of oil over high heat and deep fry fish until crispy.

5. Heat 1 tbsp of oil, stir in ginger and garlic until pungent, add sugar, fish sauce and 2 tbsp of water and cook until sauce thickens.Stir in fish with wine sprinkled along the side of the wok and mix until sauce thickens and adheres to fish.

TIPS

1. Grenadier anchovies are small fish, and should be deep fried over high heat to become crispy.

2. The fish is available only during the spring and can be purchased in the fish stalls in the wet markets.

INGREDIENTS

1 duck, about 1500 g
3 tbsp salt
1 tsp Sichuan pepper, crushed

INGREDIENTS FOR BRINE

2 star anise
1 tsp licorice root powder
2 caoguo
1 tbsp Sichuan pepper
1 section aged tangerine peel
6 cloves
6 bay leaves
1 tsp white peppercorns
20 g ginger, sliced
150 g salt
3 tbsp Shaoxing wine

METHOD

1. Cut off duck tail. Rinse duck and hang up to dry.

2. Heat 3 tbsp of salt and mix with crushed Sichuan pepper. Rub duck inside and out with the mixture and marinate for 6 hours.

3. Put all the spices for the brine in a spice pouch.

4. Put spice pouch, ginger and 2500 ml of water in a large pot, bring to a boil, reduce to low heat and simmer for 20 minutes.

5. Add 150 g of salt and 3 tbsp of wine to the brine and stir until salt is dissolved.

6. Put duck into the brine, breast face down, and bring to a boil. Reduce to low heat and allow to simmer for 30 minutes.

7. Turn duck over with the breast facing up, and cook for another 30 minutes.

8. Remove duck from the brine and set aside to cool. Cut duck to pieces and transfer to a plate.

鹽
水
鴨

SALT WATER DUCK

*Refer to p.42-44

TIPS

1. Duck tail should be cut off to avoid the gamy taste.

2. Cooking time for the duck should be adjusted based on the size.

紅燒獅子頭

BRAISED MEATBALL, HANGZHOU STYLE

Refer to p.46-49

INGREDIENTS

300 g minced pork, lean
200 g pork, fatback
1 tbsp oat meal
1 tbsp ginger juice
1/2 tsp salt
1/2 tsp sugar
1/4 tsp ground white pepper
1 tbsp corn starch

SAUCE INGREDIENTS

6 dried black mushrooms
6 sliced ginger
1 tbsp Shaoxing wine
1 tbsp red braising soy sauce
1/2 tsp sugar
1 tbsp oyster sauce
125 ml chicken broth/ water
1 tsp sesame oil
1 tsp corn starch

METHOD

1. Chop half of the minced pork further into a finer consistency, and mix with the other half.

2. Cut fatty pork into thin strips and then into tiny bits.

3. Crush oat meal into powder, soften dried mushrooms in water and remove stems. Save mushroom water for later use.

4. Put pork, ginger juice, salt, sugar, white pepper, corn starch and crushed oat meal together into a large bowl and mix by hand until all the ingredients are totally blended into a thick meat patty. Pick up the meat patty and smash it against the bowl several times to increase glutinousness. Refrigerate for 2 hours.

5. Separate meat patty into six portions, use wet hands to form 6 meatballs and coat with a thin layer of corn starch.

6. Heat frying oil to medium temperature (about 150°C), deep fry meatballs until the surfaces are firm, reduce to low heat and deep fry until meatballs are about 80% done.

7. Pour out oil, leaving only 1 tablespoonful in the wok and stir fry ginger and mushrooms until pungent, sprinkle wine and add soy sauce, oyster sauce and sugar. Put in meatballs and add chicken broth and mushroom water, bring to a boil and braise for 2 to 3 minutes. Put meatballs on a plate.

8. Reduce the sauce with high heat and thicken with corn starch. Stir in sesame oil before putting on the meatballs.

9. Boiled vegetables may be served as part of this dish.

TIPS

1. 8 smaller meatballs instead of 6 can be formed to accommodate more people.

2. Oat meal can absorb part of the juice in the meatball and increase the gumminess.

醃
篤
鮮

YANDUXIAN
SOUP

Refer to p.50-52

INGREDIENTS

200 g salted pork
200 g pork belly
30 g ginger
2 stalks spring onion
150 g Shanghai brassica
2 spring/winter bamboo shoots
200 g knotted baiye
1/2 tsp baking soda

METHOD

1. Blanch salted pork and pork belly for 3 minutes, rinse.

2. Slice ginger and knot spring onions. Trim and keep only the tender heart of the Shanghai brassica, blanch and rinse with cold water immediately to maintain their bright green color.

3. Slit open the bamboo shoots and remove the skins and cut off the hard part at the bottom shoots. Remove the hollow part of the shoots, and roll cut the remainder into chunks. Blanch for 5 minutes and drain.

4. Soak knotted baiye in 500 ml of water with 1/2 tsp of baking soda for 15 minutes. Rinse thoroughly and squeeze excess water to rid of all baking soda taste.

5. Place salted pork, pork belly, ginger and spring onion in a casserole, and add water to cover all ingredients. Bring to a boil, reduce to low heat and cook for 90 minutes.

6. Add bamboo shoots and cook for another 45 minutes.

7. Discard ginger and spring onions, cut salted pork and pork belly into pieces and return to the soup together with baiye. Boil another 5 minutes.

8. Add Shanghai brassica, sample and flavor with salt if necessary.

TIPS

1. Blanch bamboo shoots to remove any bitterness.

2. Taste soup before adding salt as much of the saltiness comes from the salted pork.

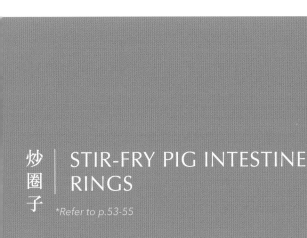

炒
圈
子

STIR-FRY PIG INTESTINE
RINGS

*Refer to p.53-55

INGREDIENTS

4 fresh pig intestines
1 tbsp corn starch
1 tbsp salt
4 tbsp white vinegar
30 g ginger slices
1 tbsp white peppercorn, crushed
2 tbsp sugar
20 g ginger, shredded
2 tbsp Shaoxing wine
2 tbsp red braising soy sauce
1 tbsp spring onion oil (see P.165)

METHOD

1. Cut off and discard about 5 cm from the thick end of the intestines, and also remove the thin end, keeping about 50 cm of the thicker part. Hold the intestine under the faucet, turn intestine inside out about 10 cm to make a cup and use running water to help push the remainder through the intestine. Rub with corn starch, rinse thoroughly. Rub with salt and rinse again. Turn the intestine inside out using the same method. Cut each intestine in half.

2. Add intestines to a large pot together with 2 tbsp vinegar, crushed white peppercorn, ginger and enough water to cover, boil for 15 minutes, reduce to medium low heat and continue to cook for about one hour, change water, add 2 tbsp vinegar and boil for 1/2 hour or until tender. Rinse, drain, and cut intestines into rings of about 1.5 cm.

3. Prepare caramel syrup with 2 tbsp of sugar (see P.165).

4. Heat 1 tbsp of oil in a wok, stir-fry ginger until pungent, put in intestine rings and stir-fry over high heat, add wine, soy sauce and caramel syrup, and continue to stir-fry until intestine rings take on slightly burnt color.

5. Stir in spring onion oil. Garnish before serving.

TIPS

Pig intestines must be very fresh for this dish to get the best results. Intestines spoil easily and should be cleaned and processed immediately to prevent spoiling.

SAUTÉED WHEAT GLUTEN WITH VEGETABLES

油麵筋塞肉燴青菜

Refer to p.56-57

INGREDIENTS

200 g minced pork
8 wheat glutens
1 tsp light soy sauce
1 tsp sugar
1 tsp corn starch
1 tsp sesame oil
1 stalk spring onion, sectioned
1 tbsp dark soy sauce
1/4 tsp salt
1/2 tsp sugar
200 g Shanghai brassica

METHOD

1. Chop minced pork to a finer texture, mix in 1/2 tsp light soy sauce, sugar, corn starch and 1 tbsp of water, and stir in sesame oil.

2. Cut an opening in each gluten and stuff with minced pork.

3. Rinse and blanch vegetables, drain.

4. Heat 2 tbsp oil, stir-fry spring onion until pungent, add gluten, meat side down, and pan fry to brown the meat. Add 1/2 tsp light soy sauce, dark soy sauce, salt, sugar and 250 ml water, and bring to a boil. Reduce to low heat and sauté for 5 minutes, add vegetable and cook another 5 minutes. Thicken sauce with corn starch.

炒蝦腰

STIR-FRY PRAWNS AND PORK KIDNEY

Refer to p.58-61

INGREDIENTS

1 pc pork kidney (about 150 g)
300 g fresh prawns
1 tbsp white vinegar
1 tsp salt
1 tbsp soy sauce
1/2 tsp sugar
1/2 tbsp Shaoxing wine
1/2 tsp corn starch
1 tbsp ginger juice
20 g ginger slices
2 stalks spring onion stem, sectioned
1/2 tsp sesame oil

METHOD

1. Slice kidney sideways into two halves, cut off the white glands, and carve out all parts that show a dark red color.

2. Soak kidney in 500 ml fresh water for 30 minutes with 1 tbsp of vinegar added. Rinse with cold water.

3. Score kidney on the surface at 3 mm interval with a depth about half the thickness of the kidney, and cut into 3 cm wide pieces. Soak kidney in fresh water until ready to cook. Replace water with fresh water once or twice.

4. Shell prawns, cut open at the back, devein, rub with 1 tsp of salt, rinse and drain.

5. Mix soy sauce, sugar, wine, corn starch and 1 tbsp water into a seasoning sauce. Stir again before use.

6. Boil 1 litre of water in a wok, add 1 tbsp of ginger juice, and put in kidney. When water re-boils, remove kidney immediately to a colander to drain.

7. Heat 2 tbsp of oil. Stir fry ginger and spring onions, add prawns and cook until about half done, stir in kidneys and seasoning sauce, reduce over high heat, add sesame oil before serving.

TIPS

1. *Select only kidney and liver that are firm and full, and shiny on the surface.*
2. *Scoring kidney helps to evenly cook the kidney.*
3. *Soaking in vinegar is the best way to clean kidney and liver.*

TOFU ROLL IN CHICKEN SOUP

Refer to p.62-64

INGREDIENTS

5 sheets tofu wrap
250 g ground pork
50 g fresh shrimps
200 g Chinese cabbage
1/2 tsp salt
1/2 tsp sugar
1 tsp light soy sauce
1 tbsp corn starch
500 ml unsalted chicken broth
1 tsp baking soda
1 bunch Chinese leeks
ground white pepper as needed

METHOD

1. Mix 1 tsp baking soda with 1 litre of warm water, immerse tofu sheets for about 15 minutes until tofu sheets change to an off white color. Rinse thoroughly to rid off the taste of baking soda.

2. Rinse, cut and blanch the cabbage, and squeeze water from the cabbage. Chop cabbage to smaller pieces.

3. Shell and coarsely chop the shrimps.

4. Mix ground pork, shrimp, cabbage, salt, sugar, soy sauce, ground white pepper and corn starch into a stuffing.

5. Quarter each tofu sheet and make tofu roll with 1 tbsp of stuffing each.

6. Cut leaves from the Chinese leeks, and run each leaf rapidly through boiling water. Cut each leaf in two lengthwise down the center.

7. Tie each tofu roll with a piece of leek leaf.

8. Bring chicken broth to a boil, put in tofu roll, reduce to low heat and cook for 10 minutes. Flavor with salt as needed.

9. Serve together with soup.

草生 | STIR-FRY ALFALFA
頭煸

Refer to p.65-67

INGREDIENTS
600 g alfalfa
1 tsp sugar
1/2 tsp salt
1 tsp light soy sauce
2 tbsp lard
1 tbsp oil
2 slices ginger
1 tbsp gaoliang wine

METHOD
1. Rinse alfalfa, drain thoroughly. Do not soak.

2. Mix sugar, salt and soy sauce in a bowl and stir until dissolved into a seasoning sauce.

3. Heat lard and oil over high heat and stir-fry ginger until pungent.

4. Put in alfalfa, add seasoning sauce and stir-fry rapidly for about 10 seconds. Transfer to plate and spray with gaoliang wine immediately.

結豆雪 | PRESERVED POTHERB MUSTARD,
百菜 | SOY BEANS AND BAIYE KNOTS
頁毛

Refer to p.68-69

INGREDIENTS
10 baiye knots
200 g preserved potherb mustard
250 g podded soy beans
10 g ginger, shredded
1/2 tsp sugar
60 ml chicken broth
1 tsp bicarbonate of soda

METHOD
1. Dissolve bicarbonate of soda in 1 litre of warm water, add baiye knots, and soak for 15 minutes until color changes to white. Rinse thoroughly and blanch for 3 minutes.

2. Soak preserved potherb mustard for 10 minutes, squeeze to remove water, and chop.

3. Blanch soy beans for 5 minutes.

4. Heat 2 tbsp oil, stir-fry ginger over high heat, add preserved potherb mustard, soy beans and sugar, and stir-fry for about 1 minute. Add chicken broth and baiye knots, and cook until sauce is reduced.

莧豆 菜瓣 | AMARANTH GREENS WITH BROAD BEANS
Refer to p.70-71

INGREDIENTS
150 g frozen broad beans
600 g red/green amaranth
2 cloves garlic, sliced
1/2 tsp salt

METHOD
1. Blanch broad beans for 5 minutes.
2. Rinse amaranth, remove roots, and cut into 2 halves.
3. Heat 2 tbsp oil, stir-fry garlic until pungent, put in broad beans and amaranth, and stir-fry over high heat until amaranth softens. Flavor with salt.

雞松 米子 | CHICKEN WITH PINE NUTS
Refer to p.72-74

INGREDIENTS
200 g boned chicken thigh, skinless
50 g pine nuts
1/2 tsp salt
1 tsp corn starch
1 tsp Shanghai soy sauce
1/2 tsp black vinegar
1 tsp sugar
1/4 tsp sesame oil
1 tbsp ginger, chopped
2 cloves garlic, coarsely chopped
1 tsp Shaoxing wine

METHOD
1. Remove tendons and ligaments from the chicken thigh, rinse, chop with the back of the knife, and dice. Mix with salt and 1/2 tsp corn starch.
2. Mix soy sauce, vinegar, sugar, sesame oil, 1/2 tsp corn starch and 1 tbsp water into a seasoning sauce. Stir again before use.
3. Roast pine nuts in a dry wok.
4. Heat 2 tbsp oil in a wok, stir-fry ginger until pungent, add chicken and garlic and stir-fry until about 80% done.
5. Drizzle wine, add pine nuts, put in seasoning sauce and sauté until thickened.

TIPS

Pre-mix seasoning sauce is to reduce the cooking time of chicken in the wok to get the best result.

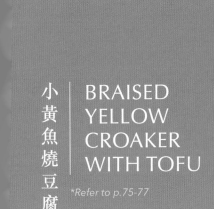

小黃魚燒豆腐 | BRAISED
YELLOW
CROAKER
WITH TOFU

*Refer to p.75-77

INGREDIENTS

450 g small yellow croaker
250 g tofu
1 tsp salt
10 g ginger slices
1 stalk spring onion, sectioned
2 tbsp Shanghai soy sauce
1/2 tsp sugar
1 tbsp Shaoxing wine
1 tsp Zhenjiang vinegar
1 tbsp corn starch

METHOD

1. Clean the fish of scales and gills, make a cut at the end of the fish stomach, insert two chopsticks through the head into the stomach, grab hold of the fish maw and intestines, twist and pull out intestines, then flush the inside of the stomach thoroughly.

2. Coat fish with a thin layer of corn starch and pan fry until brown.

3. Cut tofu into thick pieces, soak in 500 ml of warm water and salt for 15 minutes, drain, and pan fry until golden.

4. Mix soy sauce, sugar, vinegar, corn starch and 250 ml water into a seasoning sauce.

5. Heat 2 tbsp oil in a casserole and stir-fry ginger and spring onion until pungent. Line the bottom of the casserole with tofu with the fish on top, add wine, put in seasoning sauce and boil for about 1 to 2 minutes.

蝦毛仁豆 | SHRIMPS WITH PODDED SOY BEANS
Refer to p.78-79

INGREDIENTS
300 g shelled fresh water shrimps
150 g podded soy beans
1/4 tsp salt
1/2 egg white
1 tsp corn starch
10 g ginger slices
1 tbsp Shaoxing wine

TIPS

1. *The best way to thaw frozen shrimp is to soak it in iced water and allow them to thaw slowly.*

2. *Frozen shelled fresh water shrimps are available in shops selling Shanghainese foods. If shelled fresh water shrimps are not available, frozen shelled white shrimps from South America can be used as a replacement.*

METHOD
1. Rinse shrimps and drain. Put shrimps into a bowl and gently rub with 1/4 tsp of salt until slightly gummy, add egg white and mix, and stir in corn starch.

2. Boil soy beans for 5 to 6 minutes, drain.

3. Heat 500 ml of oil to a low temperature (about 120°C), add shrimps, disperse with chopsticks for about 15 seconds and remove immediately to drain. Pour out oil leaving only 1 tablespoonful in the wok.

4. Heat the oil in the wok over high heat, stir fry ginger, add soy beans and shrimps and toss until shrimps are fully cooked, drizzle wine along the inside of the wok, and stir fry rapidly until all the wine evaporates.

巴紅甩燒水下 | BRAISED FISH HEAD AND TAIL
Refer to p.80-82

INGREDIENTS
1 fish head
1 fish tail
1/2 tbsp salt
3 tbsp corn starch
2 stalks spring onion, sectioned
2 tbsp Shaoxing wine
1 tbsp ginger juice
2 tbsp red braising soy sauce
2 tbsp sugar

METHOD
1. Clean fish head and cut into two halves, rinse and drain. Clean fish tail and cut vertically into 2 or 3 pieces. Marinate fish head and tail with salt for about 15 minutes, and mix in 2 tbsp of corn starch.

2. Heat 250 ml oil and brown both sides of the fish head and tail.

3. Heat 2 tbsp oil in the wok, stir-fry spring onions until pungent, put in fish head and tail, drizzle wine, add ginger juice, soy sauce, sugar and 250 ml of water, and bring to a boil. Reduce to low heat, cover, and simmer about 5 minutes, remove cover, and reduce sauce over high heat.

4. Transfer fish head and tail to a plate, mix 1 tbsp corn starch with water and thicken sauce, and drizzle over fish.

響油鱔糊 | SAUTÉED EEL

*Refer to p.83-85

INGREDIENTS

500 g rice field eel
1.5 tbsp corn starch
1 tsp salt (for cleaning)
2 tbsp Shanghai soy sauce
1 tsp sugar
1.5 tbsp lard
1 tbsp shredded ginger
1 tsp Shaoxing wine
100 g hotbed chives, sectioned
1/2 tsp ground white pepper
2 tbsp chopped garlic
1 tbsp sesame oil

METHOD

1. De-bone eel, discard head and bones, and remove intestines. Clean the eel with 1/2 tsp salt and 1 tbsp corn starch, rinse and drain, and cut into sections 8 to 10 cm long and pieces of about 1 cm wide.

2. Bring a large pot of water to a boil, and turn off the heat. Put in eel pieces and stir rapidly for about 10 seconds, then rinse with cold water and drain.

3. In a small bowl, mix 1/2 tsp salt, 1/2 tbsp corn starch, soy sauce, sugar and 1 tbsp water into a sauce. Stir again before use.

4. Heat up the lard in the wok, put in shredded ginger and eel pieces and stir-fry rapidly to rid of the moisture. Sprinkle wine along the inside of wok, add the sauce and stir-fry rapidly until eel pieces are thoroughly cooked, add hotbed chives and toss, put in ground white pepper, mix well and transfer to a plate.

5. Move the eel pieces in the plate to create an opening in the center and add garlic.

6. In a clean wok, heat up 1 tbsp of oil and 1 tbsp of sesame oil; pour over the chopped garlic before serving.

TIPS

1. Sautéed eel has to be done very rapidly over high heat to avoid overcooking. Sauces should be prepared in advance.

2. Do not overcook hotbed chives as they cook easily. Overcooking will result in sogginess.

毛油蟹醬 | SAUTÉED RIVER CRAB IN SOY SAUCE

Refer to p.86-87

INGREDIENTS

4 fresh water crabs (small)
150 g podded soy beans
2 tbsp flour
1 tbsp corn starch
1 stalk spring onion, chopped
20 g ginger, chopped
2 tbsp Shaoxing wine
1 tbsp sugar
2 tbsp Shanghai soy sauce
1 tsp corn starch (for thickening)
1 tbsp lard

METHOD

1. Remove carapace of each crab, and cut crab into two halves, remove stomach and gills, rinse.
2. Blanch soy beans.
3. Mix flour and corn starch, and dust the cut end of each crab. Heat 2 tbsp oil and pan fry the cut ends of each crab.
4. Heat 1 tbsp oil and stir-fry ginger and spring onion until pungent, put in crabs and toss. Drizzle wine, add soy beans, sugar, soy sauce and 250 ml water, and cook until crabs are fully cooked. Thicken sauce with corn starch, add lard and toss thoroughly.

大墨 燆魚 | STEWED PORK BELLY WITH CUTTLEFISH
Refer to p.88-89

INGREDIENTS
600 g pork belly
300 g fresh cuttlefish
4 eggs
2 tbsp sugar (for caramel)
50 g ginger slices
1/2 tbsp Shanghai doubanjiang
3 tbsp Shaoxing wine
2 star anise
2 tbsp red braise soy sauce
1 tsp salt
40 g rock sugar

METHOD
1. Blanch pork belly and cut into 3 cm chunks. Clean cuttlefish and cut into 4 cm squares.
2. Boil eggs, soak in cold water until cool, and peel. Score the eggs vertically 5 or 6 times with each cut penetrating the egg yolk.
3. Prepare caramel (see P.165).
4. Heat 2 tbsp oil and stir-fry ginger and doubanjiang, put in pork belly and toss, add wine, star anise, caramel and enough water to cover pork, bring to a boil, reduce to low heat, cover and simmer for about 1 hour.
5. Add soy sauce, salt, rock sugar and cuttlefish, and continue to simmer for 15 minutes. Add eggs, simmer for another 5 minutes, and reduce sauce over high heat.

跑銀 蛋魚 | SCRAMBLED EGGS WITH NOODLE FISH
Refer to p.90-91

INGREDIENTS
350 g noodle fish
2 eggs
1 spring onion, shredded
1/4 tsp salt
1/8 tsp ground white pepper
2 tbsp lard

METHOD
1. Rinse fish and blanch for 20 seconds. Drain and absorb excess water with kitchen towels.
2. Beat eggs, and mix in fish, spring onion, salt and ground white pepper into a batter.
3. Heat lard over high heat, add batter and stir-fry rapidly with chopsticks for about 8 to 10 seconds until eggs coagulated.

茄醬 子爆 | SAUTÉED EGGPLANT WITH THICK SAUCE
*Refer to p.92-93

INGREDIENTS
600 g egg plant
1 tsp salt
2 tbsp Shanghai doubanjiang
1 tbsp sugar
2 tbsp ginger, chopped
1 stalk spring onion, sectioned
1 stalk spring onion, chopped
1 tbsp garlic, grated
2 tbsp Shaoxing wine

METHOD
1. Clean egg plants, and roll cut into chunks. Soak in salted water. Drain just before cooking.
2. Mix doubanjiang with sugar and 2 tbsp water into a seasoning sauce.
3. Heat 3 tbsp oil, stir-fry ginger, spring onion sections and garlic until pungent, put in egg plants and stir-fry until slightly brown. Add wine, seasoning sauce, and sauté until the sauce thickens. Toss to mix well, add chopped spring onion before plating.

麵雙 筋菇 | BRAISED WHEAT GLUTEN WITH MUSHROOMS
*Refer to p.94-95

INGREDIENTS
10 wheat glutens
4 dried black mushrooms
8 small button mushrooms
2 spring onion stems, sectioned
1 tsp sugar
1 tsp light soy sauce
1/2 tbsp dark soy sauce
1 tbsp corn starch
1 tsp sesame oil

METHOD
1. Rinse and soak black mushrooms until soft, remove stems, and cut in half. Rinse button mushrooms.
2. Heat 2 tbsp oil in a wok, stir-fry spring onions until pungent, add mushrooms and enough water to cover, bring to a boil, add wheat glutens, sugar and the soy sauces, cover, reduce to low heat and braise for 5 minutes. Reduce sauce and thicken with corn starch. Stir in sesame oil.

魚糟 片溜 | FISH IN DISTILLED GRAIN SAUCE

Refer to p.96-98

INGREDIENTS

250 g flounder fillet
1/2 tsp salt
1 egg white
3 tbsp corn starch
5 g dried black fungus
50 g bamboo shoot slices
10 g ginger slices
4 tbsp distilled grain sauce
1 tbsp Shaoxing wine
2 tsp sugar
2 tbsp spring onion oil
(see P.165)

METHOD

1. Slant cut fish fillet into 3 mm thick slices, add salt and mix gently by hand until gummy, put in egg white and mix thoroughly, and blend in 1 tbsp corn starch. Marinate for 15 minutes.

2. Soak black fungus until soft, and blanch. Blanch bamboo shoot.

3. Boil water in a wok, gently put in fish slices one at a time, turn off heat, and allow the fish to steep in hot water until cooked. Remove fish.

4. Heat 1 tbsp spring onion oil and stir-fry ginger until pungent. Add bamboo shoot, black fungus, distilled grain sauce, wine, sugar and 250 ml water, and bring to a boil. Mix 2 tbsp corn starch with water and thicken sauce in the wok. Add fish, drizzle 1 tbsp spring onion oil, stir gently and transfer to a bowl.

蝦水 仁晶 | CRYSTAL SHRIMPS

Refer to p.100-101

INGREDIENTS

300 g fresh water shrimps
1/2 tsp salt
1/2 tsp sugar
1 egg white
1 tbsp corn starch

METHOD

1. Rub shrimps with 1/4 tsp salt, rinse, and flush under running water for 5 minutes. Drain and absorb excess water with kitchen towels.

2. Mix shrimps with sugar and the remaining salt, add egg white, and mix well with corn starch. Put shrimps in a strainer to drain excess liquid.

3. Heat 500 ml oil to a medium temperature (about 140°C, or when small bubbles begin to form when a damp chopstick is inserted), add shrimps and stir gently with chopsticks to disperse. Turn off heat and cook shrimps in hot oil for about 2 minutes. Remove shrimps and drain, and transfer to a plate.

TIPS

1. The best way to thaw frozen shrimp is to soak it in iced water and allow them to thaw slowly.

2. Frozen shelled fresh water shrimps are available in shops selling Shanghainese foods. If shelled fresh water shrimps are not available, frozen shelled white shrimps from South America can be used as a replacement.

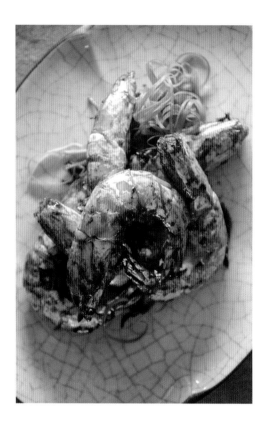

炸烹大蝦 | DEEP FRIED JUMBO PRAWNS WITH PUNGENT SAUCE

Refer to p.102-103

INGREDIENTS

600 g jumbo prawns
2 tbsp unsalted chicken broth
1 tbsp light soy sauce
1/2 tsp salt
1 tsp sugar
1 tbsp Zhenjiang vinegar
1/2 tsp corn starch
1 tbsp Shaoxing wine
2 cloves garlic, sliced
10 g ginger, shredded
2 stalks spring onion, shredded

METHOD

1. De-vein prawn by inserting a tooth pick behind the second abdominal segment to pick out and remove the vein, then cut off antenna and legs using kitchen shears.

2. Insert kitchen shears into the head and cut open the shell from head to tail along the abdomen of the prawn.

3. Using a sharp knife, cut open the body from the tail to the head, and coat with a thin layer of corn starch.

4. Mix light soy sauce, salt, sugar, vinegar, corn starch and 2 tbsp of chicken broth into a seasoning sauce.

5. Heat 500 ml oil to medium temperature (about 150°C) in a wok and deep fry prawns until about 80% done. Remove the prawns and pour out oil, leaving only 2 tbsp in the wok.

6. Stir fry garlic and ginger, add prawns, sprinkle wine, stir and add seasoning sauce. Bring to a boil, and stir-fry until sauce is reduced. Top with spring onion.

蟹粉豆腐 | # TOFU WITH CRAB MEAT AND CRAB ROE
Refer to p.110-112

INGREDIENTS

3 salted-eggs
50 g ginger
50 g carrot
2 pc tofu
1 tsp salt
1 tsp sugar
1 tsp Zhenjiang vinegar
2 tsp corn starch
3 tbsp crab meat and tomalley
200 g crab roe

METHOD

1. Steam salted eggs until fully cooked. Crush and save egg yolks. Discard egg whites.

2. Chop ginger, grate carrot, and cut tofu into 1.5 cm cubes.

3. Heat 2 tbsp of oil in the wok, and stir fry ginger and carrot until pungent. Put in 3/4 of the salted egg yolks and 60 ml of water, bring to a boil and season with salt, sugar and vinegar.

4. Drain and add tofu into the wok, stir gently and cook for about 30 seconds. Thicken with corn starch and transfer to a deep plate.

5. Add 2 to 3 tbsp of water to the wok, boil, and stir in crab meat and tomalley and the remaining salted egg yolks. Dish out to put on top of tofu.

6. Top with crab roe.

TIPS

1. Putting in salted egg yolks twice separately enhances both the flavor and color.

2. Vinegar heightens the flavor of the crab but too much will affect the color of the dish.

INGREDIENTS

1 soft shell turtle
5 dried black mushrooms
20 g Jinhua ham, sliced
1 tsp light soy sauce
1/2 tsp sugar
1 tsp salt
1 tbsp ginger juice
1 tbsp Shaoxing wine
1/2 tsp ground white pepper
1 tbsp corn starch
1/2 tsp sesame oil
2 spring onion stems, shredded

清蒸甲魚 | # STEAMED SOFT SHELL TURTLE

Refer to p.113-115

METHOD

1. Slaughter and rinse turtle with water, blanch with 60°C hot water and remove black film on the shell. Remove shell and cut turtle meat into pieces and soak in cold water for 15 minutes. Rinse and drain, and blanch turtle in boiling water for 1 minute. Drain. Blanch and save shell for later use.

2. Soften mushrooms in water and remove stems. Slant cut into thick pieces. Soak ham for 5 minutes, drain.

3. Marinate turtle with soy sauce, sugar, salt, ginger juice, wine and ground white pepper for 15 minutes. Mix in 1/2 tbsp corn starch and stir in sesame oil and 1 tsp oil.

4. Add mushrooms and ham, and mix well. Turn over turtle shell and use it as a receptacle for the mixed ingredients. Steam for 25 minutes and reserve the juice from inside the shell.

5. Cover turtle shell with a large plate, turn over to transfer the ingredients to the plate. Mix 1/2 tbsp corn starch with water and thicken the reserved juice. Pour over the turtle and top with shredded spring onions.

> **TIPS**
>
> *Follow the order in which ingredients and condiments are added to ensure maximum flavor.*

葱烏蝦黄參籽拌大 | BRAISED SEA CUCUMBER WITH SHRIMP ROE AND CHINESE LEEK

Refer to p.107-109

INGREDIENTS

1 large sea cucumber, pre-softened
250 ml unsalted chicken broth
2 tbsp shrimp roe
4 Chinese leeks
4 tbsp spring onion oil (see P.165)
30 g ginger slices
2 cloves garlic, sliced
2 tbsp Shaoxing wine
1 tbsp sugar
1/2 tsp salt
1 tbsp dark soy sauce
corn starch as needed

METHOD

1. Braise sea cucumber in chicken broth over low heat for 30 to 45 minutes or until sufficiently soft.

2. Roast shrimp roe in a dry wok over low heat.

3. Use only about 4 cm sections of the stem of the leeks, remove 2 to 3 outer layers to expose the heart, and pan fry in 3 tbsp of spring onion oil over low heat until slightly brown. Remove from wok.

4. Heat the remaining oil in the wok, stir-fry ginger and garlic until pungent, change to medium heat and stir in shrimp roe. Drizzle wine.

5. Add sea cucumber, salt, sugar, soy sauce and the chicken broth, bring to a boil, put in leeks, reduce to low heat and braise until sauce begin to thicken. Thicken sauce further with corn starch if needed. Finally stir in 1 tbsp spring onion oil.

燒魚肉唇紅 | STEWED PORK BELLY WITH SHARK SKIN

Refer to p.119-120

INGREDIENTS

600 g pork belly
100 g dried shark skin
50 g ginger slices
30 g sugar
1 tbsp Shanghai doubanjiang
2 star anise
4 tbsp Shaoxing wine
2 tbsp red braise soy sauce
1 tsp salt
40 g rock sugar

METHOD

1. Prepare fish skin, and cut into large pieces.

2. Blanch pork belly for 3 minutes, rinse, and cut into 3 cm thick pieces.

3. Prepare caramel (see P. 165)

4. Heat 2 tbsp oil, stir-fry ginger and doubanjiang until pungent, put in pork and stir-fry over high heat. Add star anise, wine and enough water to cover pork, bring to a boil, reduce to medium heat, cover and stew for 30 minutes.

5. Add soy sauce, salt, rock sugar and caramel, re-boil, reduce to low heat and stew for 45 minutes. Put in shark skin, and stew another 15 minutes. Reduce sauce over high heat.

丁楓 蹄涇 | BRAISED PORK KNUCKLE
Refer to p.121-123

INGREDIENTS
2 pork knuckle
3 tbsp sugar (for caramel)
3 g cloves
6 g cinnamon
30 g ginger slices
125 ml Shaoxing wine
3 tbsp red braising soy sauce
1 tsp salt
40 g rock sugar

METHOD
1. Clean knuckle and chop into two chunks. Blanch and drain.
2. Prepare caramel syrup (see P. 165).
3. Put cloves and cinnamon in a spice pouch.
4. Heat 2 tbsp of oil in a pot, stir ginger until pungent, put in spice pouch, knuckle and all the remaining ingredients, add water to cover, bring to a boil, cover, reduce to low heat and braise until knuckles are sufficiently soft. Remove spice pouch, reduce sauce over high heat.

河紅 鰻燒 | BRAISED EEL
Refer to p.124-126

INGREDIENTS
1 pc eel (about 900 g)
1 tbsp salt
50 g ginger
4 stalks spring onion
4 tbsp lard
2 tbsp Shaoxing wine
2 tbsp sugar
2 tbsp red braising soy sauce

TIPS
Select eel that is small and narrow. Large eel requires longer cooking time and the skin is more likely to come apart.

METHOD
1. Soak eel in hot water (500 ml boiling water and 250 ml cold water) for 2 minutes. Brush mucus off the skin by hand, rub with salt and wash with fresh water.
2. Cut off and discard head and tail. Select the middle portion of the eel and cut into 6 to 7 cm sections.
3. Slice ginger and section spring onions.
4. Heat 2 tbsp of lard, stir fry ginger and spring onions until pungent, and then add eel and wine.
5. Add 500 ml of water and bring to a boil over high heat. Reduce to low heat, and put in sugar and soy sauce after 15 minutes. Continue to braise and add lard in small portions. Braise for another 30 minutes or until eel is totally soft.
6. Change to high heat to reduce sauce, and stir in the remaining lard.

尾苔蝦條鳳 | FANTAILED PRAWNS WITH SEAWEED
Refer to p.128-129

INGREDIENTS
4 jumbo prawns
1 tbsp ginger juice
1 stalk spring onion, sectioned
1/4 tsp ground white pepper
1/4 tsp salt
1 tsp Shaoxing wine
2 tsp corn starch
1/2 egg white
3 tbsp bread crumbs
2 tbsp seaweed powder

METHOD
1. Remove head of the prawns. Shell prawns but leave the shell on the tails.
2. Cut open prawns from the abdomen to the back into two linked halves without severing the pieces. De-vein prawns, spread out the two halves of each prawn and press down to make the prawn into a flat sheet. Marinate prawns for 10 minutes with ginger juice, spring onions, white pepper, salt and wine. Discard spring onions and absorb excess moisture with kitchen towels.
3. Make a cut in the middle of each prawn and stuff the tail through the cut to the back.
4. Coat prawns with a thin layer of corn starch, blend in egg white, and mix with bread crumbs.
5. Heat 750 ml of oil in a wok to medium temperature (about 150°C), hold each prawn by the tail, lower into the oil and deep fry until brown. Remove prawns. Reheat oil to a medium high temperature (about 170°C), put in prawns and deep fry for about 5 seconds. Remove to drain oil. Mix with seaweed powder before serving.

年苔糕菜 | GLUTINOUS RICE CAKE WITH SEAWEED
Refer to p.130-131

INGREDIENTS
300 g glutinous rice cake
5 g dried seaweed
3 tbsp lard
1 tbsp sugar

METHOD
1. Rinse rice cakes and slant cut into 0.5 cm pieces.
2. Roast seaweed in a dry pan over very low heat and crush into powder.
3. Heat 1 litre water to about 90°C, turn off heat, and soak rice cakes for about 2 minutes. Drain and mix with 1 tbsp oil thoroughly so that each piece of rice cake is covered with oil.
4. Heat 3 tbsp lard in a wok, put in rice cakes and sugar, separate each piece with chop sticks, and stir-fry until slightly browned. Turn off heat and mix with crushed seaweed.

砂鍋餛飩雞 | WONTON AND CHICKEN CASSEROLE

*Refer to p.132-133

INGREDIENTS

1/2 chicken
50 g Jinhua ham
150 g minced pork
1 tsp light soy sauce
1/2 tsp salt
1/2 tsp sugar
300 g Chinese cabbage
1 tbsp corn starch
1/2 tbsp sesame oil
200 g Shanghainese wonton skin
250 ml chicken broth

METHOD

1. Rinse chicken, blanch and drain.

2. Slice ham, place on a plate and cover ham with water. Steam for 10 minutes and pour out water from the plate.

3. Marinate minced pork with soy sauce, salt, sugar and 2 tbsp water for 15 minutes.

4. Blanch vegetables until soft and rinse with cold water. Chop vegetables and squeeze out most of the water. Mix thoroughly with minced pork.

5. Add corn starch and stir in one direction until fully blended. Add sesame oil and mix well to make wonton filling.

6. Wrap wontons and put into a large pot of boiling water. When the water boils again, add half a cup of cold water to lower the outside temperature of the wontons. If the wontons come afloat when the water re-boils, remove wontons.

7. Heat chicken broth in a casserole, add ham and chicken and enough water to cover all the ingredients, bring to a boil, cover and turn off heat. Add wontons after 15 minutes and bring to a boil again. Salt to flavor if needed.

荷葉八寶雞

STUFFED CHICKEN WRAPPED IN LOTUS LEAF

Refer to p.134-137

INGREDIENTS

160 g glutinous rice
50 g Jinhua ham, diced
12 shelled chestnuts
1 dressed chicken
1.5 tbsp salt
1 tbsp ginger juice
1 tbsp Shaoxing wine
2 pc dried lotus leaf
2-3 dried black mushrooms
30 g dried scallops
20 g dried shrimps
30 g lotus seeds (core removed)
20 g shelled raw peanuts
1 tbsp dark soy sauce

METHOD

1. Soak glutinous rice in hot water with 1/2 tbsp salt for 1 hour, rinse and drain.

2. Rinse ham and steam for 10 minutes.

3. Remove husk from the chestnuts, and steam for 1 hour.

4. Clean chicken, debone, and marinate with 2 tsp of salt, 1 tbsp ginger juice and 1 tbsp wine for 30 minutes.

5. Rinse lotus leaf and soak until soft.

6. Soak mushrooms until soft, remove stems, and dice.

7. Soak dried scallops until soft, and tear into shreds. Save the water for later use.

8. Soak dried shrimps until soft. Drain before use.

9. Rinse lotus seeds and peanuts, and steam for 15 minutes.

10. Stir-fry ham, mushrooms, scallops, dried shrimps, peanuts, chestnuts and lotus seeds in 2 tbsp of oil for about 1 minute, add rice, 1 tbsp dark soy sauce and water from soaking scallops. Mix well and transfer to a large plate. Steam for about 45 minutes or until rice is fully cooked. Set aside to cool.

11. Stuff chicken with glutinous rice, wrap in lotus leaf, and steam over high heat for 30 minutes.

素翡 方翠 | VEGETABLE ROLL

Refer to p.138-140

INGREDIENTS

1 sheet tofu skin
450 g spinach
4 dried black mushrooms
1 tbsp porcini powder
1 egg
1 tbsp sesame
3/4 tsp salt
1/2 tsp sesame oil

TIPS

1. *For people with gout, spinach can be replaced by xiancai (amaranth greens) or other leafy green vegetables. Shredded carrots can also be added.*

2. *Porcini powder can enrich the flavor of the vegetables.*

METHOD

1. Blanch spinach and rinse immediately with cold drinking water, squeeze out excess water and remove about 4 cm stems, and coarsely chop remaining leaves and tender stems

2. Soften mushrooms with water, remove stems, and cut into thin strips. Beat egg and roast sesame.

3. Mix spinach, mushrooms, porcini powder, sesame, salt and sesame oil, and separate into four equal portions.

4. Fold tofu skin into a quadrant, cut off the round edges to form a square, unfold the tofu skin and cut into four equal pieces.

5. Brush a tofu skin with egg, put in a portion of fillings, and roll into a flat roll. Repeat with the remaining tofu skins.

6. Heat 1 tbsp oil in a pan, brown vegetable rolls on both sides over low heat, absorb excess oil with kitchen towels, and cut each roll into four pieces before serving.

粉油 絲豆 湯腐 | TOFU PUFF AND MUNG BEAN VERMICELLI SOUP

Refer to p.142-143

INGREDIENTS

150 g spareribs
8 tofu puff
20 g mung bean vermicelli
30 g Sichuan preserved mustard
1 tsp salt
1/8 tsp ground white pepper

METHOD

1. Blanch spareribs, rinse. Bring to a boil 1.5 litres of water, add spareribs and cook boil for 30 minutes into a meat broth.

2. Cut each tofu puff into two halves, and boil in light salty water for 5 minutes. Drain.

3. Soak vermicelli for 5 minutes. Drain.

4. Rinse Sichuan preserved mustard, and cut into thin slices.

5. Boil meat broth, put in tofu puffs and Sichuan preserved mustard and cook for 5 minutes. Add vermicelli and cook for another 3 minutes. Flavor with salt and ground white pepper.

羹仁海豆參腐蝦 | SEA CUCUMBER, SHRIMPS AND TOFU SOUP
*Refer to p.144-145

INGREDIENTS
250 g sea cucumber, pre-softened
250 g soft tofu
100 g shelled shrimps
250 ml unsalted chicken broth
1 tbsp ginger juice
1/8 tsp ground white pepper
1/2 tsp salt
2 tbsp water chestnut powder
1 egg white

METHOD
1. Cut sea cucumber into 1 cm cubes, blanch and drain.
2. Rinse shrimps, and cut into the same size.
3. Cut tofu into the same size.
4. Bring to a boil chicken broth, add ginger juice, sea cucumber, tofu and shrimps, boil for about 3 minutes, flavor with ground white pepper and salt, and thicken with water chestnut powder mixed with water. Turn off heat.
5. Beat egg white, and add to the soup through a sift. Stir to blend after 1 minute.

丸薺湯菜肉 | SHEPHERD'S PURSE AND MEATBALL SOUP
*Refer to p.146-147

INGREDIENTS
600 g shepherd's purse
200 g minced pork
1 tsp light soy sauce
1 tsp sugar
1 tsp corn starch
2 eggs
500 ml chicken broth
1 tsp lard
1/2 tsp salt

METHOD
1. Rinse shepherd's purse and cut off the roots.
2. Blanch shepherd's purse, rinse and drain.
3. Stir fry shepherd's purse in 1 tbsp of oil over medium heat, then put into a blender together with the chicken broth and blend into a shepherd's purse soup.
4. Mix pork, soy sauce, sugar and corn starch, add 2 tbsp water, blend in half an egg white, and stir in a single direction until gummy. Use wet hands make meatballs.
5. Boil shepherd's purse soup, put in meatballs and cook for about 2 minutes, add lard and salt, stir, turn off heat, beat eggs and drizzle evenly into the soup.

酸辣湯 | HOT AND SOUR SOUP

Refer to p.148-150

INGREDIENTS

100 g cooked pork tripe
200 g coagulated pig's blood
4 dried black mushrooms
5 g dried black fungus
1/2 pc (125 g) firm tofu
1 bamboo shoot
1 egg, beaten
1/2 tbsp salt
1/2 tsp sugar
2 tbsp water chestnut starch
5 tbsp Zhenjiang vinegar
1 tbsp ground white pepper
1 tsp sesame oil

METHOD

1. Cut pork tripe into strips. Cut pig's blood into strips and blanch.

2. Soften mushrooms and black fungus in water, and cut into thin strips.

3. Soak tofu in water with 1/4 tsp of salt added for 15 minutes, drain and cut into thin strips.

4. Remove outer layers of the bamboo shoot, shave the surface, and cut into 4 cm chunks. Cut each section into two halves and cut into thin slices, and cut into thin strips. Blanch bamboo shoot with 1/2 tsp of salt for 2 minutes, drain.

5. Boil 1 litre of water in a pot, put in pork tripe, mushroom, black fungus, bamboo shoot and pig's blood, add sugar and 1 tsp of salt, re-boil, and disperse ingredients gently with chopsticks (do not stir).

6. Mix water chestnut starch with a small amount of water until it becomes a wet starch, drizzle it into the soup slowly and stir gently.

7. Add tofu, re-boil, reduce to low heat, pour egg gradually into the soup through a sieve. Do not stir. Turn off heat.

8. Put in vinegar and ground white pepper, mix, and stir in sesame oil.

TIPS

1. Black vinegar such Laochen vinegar or Zhenjiang vinegar would bring the best results. White vinegar can also be used but it gives off a sharper taste.

2. Pork tripe can be omitted for those not eating offal,

3. Canned bamboo shoot can be used in place of fresh bamboo shoot. Blanching can help to remove its tangy taste.

葱油拌麵 | NOODLES WITH SPRING ONION OIL

Refer to p.152-153

INGREDIENTS

100 g Shanghai thin noodles
2 tbsp chicken broth
1 tsp light soy sauce
1 tsp dark soy sauce
deep fried spring onion, as needed
3 tbsp spring onion oil (see P.165)

METHOD

1. Cook noodles, drain and put into a bowl.

2. Mix chicken broth and soy sauces into a seasoning sauce and add to the noodles. Top with deep fried spring onions.

3. Heat spring onion oil and add to the noodles. Mix well.

粢飯 | SHANGHAI GLUTINOUS RICE ROLL

Refer to p.154-155

INGREDIENTS

600 g glutinous rice
150 g Sichuan preserved mustard
1 pc fried dough
1/2 cup pork floss
1 tsp sugar

METHOD

1. Soak glutinous rice in hot water for 1 hour, drain, wash away some of the starch with cold water, put in a deep plate and steam for about 45 minutes until done. Stir with a spoon to air the rice, and separate into 4 portions.

2. Soak Sichuan preserved mustard in cold water for 5 minutes, chop, stir-fry in 1 tsp of oil until pungent and mix with sugar.

3. Cut fried dough into four pieces.

4. Spread a piece of slightly oiled cling wrap atop a clean cloth on a gloved hand, put in a portion of the glutinous rice, flatten slightly with a spoon, and put in 1 piece of fried dough together with some preserved mustard and pork floss. Use the cloth and cling wrap to roll the contents into a thick rice roll.

葱開煨麵 BRAISED NOODLES WITH DRIED SHRIMPS AND SPRING ONION OIL

Refer to p.156-157

INGREDIENTS
300 g Shanghai noodles
50 g dried shrimps
2 tbsp spring onion oil (see P.165)
1 tbsp Shaoxing wine
250 ml unsalted chicken broth
as needed salt

METHOD
1. Rinse dried shrimps, and soak for 5 minutes. Drain.
2. Cook noodles, drain.
3. Heat spring onion oil, stir-fry dried shrimps, add wine, chicken broth and 250 ml of water, and bring to a boil. Put in noodles and braise over low heat for 5 to 6 minutes. Flavor with salt.

上海粗炒 STIR-FRY SHANGHAI THICK NOODLES

Refer to p.158-160

INGREDIENTS
300 g Shanghai thick noodles (dry)
3 dried black mushrooms
300 g cabbage
150 g lean pork
1/2 tsp salt
1.5 tsp sugar
1/2 tsp corn starch
1/2 tbsp light soy sauce
1.5 tbsp dark soy sauce
1 tsp sesame oil

METHOD
1. Cook noodles in a large pot of water for about 12 minutes or until cooked, rinse with cold water and drain.
2. Soak mushrooms until soft, remove stems, and cut into thin strips. Cut cabbage into small pieces, rinse and drain.
3. Cut pork into strips and marinate with 1/2 tsp each of salt, sugar, and corn starch and 1 tbsp of oil.
4. Mix the soy sauces, 1 tsp of sugar and 200 ml of water or unsalted chicken broth into seasoning sauce.
5. Heat 2 tbsp of oil, add pork, cabbage, mushrooms and noodles together with the seasoning sauce, bring to a boil, cover, and simmer over low heat for 8 or 10 minutes, then toss until all ingredients are blended. Stir in sesame oil and transfer to plate.

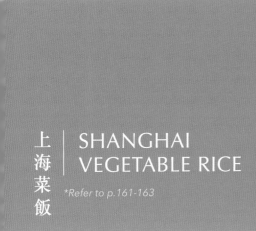

上海菜飯 | SHANGHAI VEGETABLE RICE

Refer to p.161-163

INGREDIENTS

320 g white rice
100 g Shanghai salted pork
100 g Shanghai brassica
30 g pork fatback
1 tbsp chopped garlic
1/2 tsp salt

METHOD

1. Wash rice and drain.

2. Soak the Shanghai salted pork in cold water for 1 hour, steam for 10 minutes and cut into 1/2 cm thin slices.

3. Rinse the Shanghai brassica, blanch rapidly, drain and chop into small pieces.

4. Dice pork fatback into small bits.

5. Put pork fatback bits together with 2 tbsp of water into the wok and cook until fatty pork bits become crispy. Remove and save the pork crisp for later.

6. Add chopped garlic to the oil, stir fry until pungent and put in rice, continue to stir fry for half minute.

7. Add salt and cook rice normally in a rice cooker.

8. When rice starts to boil, put in Shanghai salted pork slices.

9. When rice is cooked, add chopped Shanghai brassica, mix well, and cover the lid for 3 minutes. Sprinkle pork crisp on top before serving.

市面上的食譜書，包括我們陳家廚坊系列，食譜中的計量單位，都是採用公制，即重量以克來表示，長度以厘米 cm 來表示，而容量單位以毫升 ml 來表示。世界上大多數國家都採用公制，但亦有少數國家如美國，至今仍使用英制（安士、磅、英吋、英呎）。

香港和澳門，一般街市仍沿用司馬秤（斤、兩），在香港超市則有時用公制，有時會用美制，所以香港是世界上計量單位最混亂的城市，很容易會產生誤會。與香港關係緊密的中國大陸，他們的大超市有採用公制，但一般市民用的是市制斤兩，這個斤與兩，實際重量又與香港人用的司馬秤不同。

鑒於換算之不方便，曾有讀者要求我們在食譜中寫上公制及司馬秤兩種單位，但由於編輯排版困難，實在難以做到。考慮到實際情況的需要，我們覺得有必要把度量衡的換算，以圖表方式來說清楚。

重量換算速查表 （公制換其他重量單位）

克	司馬兩	司馬斤	安士	磅	市斤
1	0.027	0.002	0.035	0.002	0.002
2	0.053	0.003	0.071	0.004	0.004
3	0.080	0.005	0.106	0.007	0.006
4	0.107	0.007	0.141	0.009	0.008
5	0.133	0.008	0.176	0.011	0.010
10	0.267	0.017	0.353	0.022	0.020
15	0.400	0.025	0.529	0.033	0.030
20	0.533	0.033	0.705	0.044	0.040
25	0.667	0.042	0.882	0.055	0.050
30	0.800	0.050	1.058	0.066	0.060
40	1.067	0.067	1.411	0.088	0.080
50	1.334	0.084	1.764	0.111	0.100
60	1.600	0.100	2.116	0.133	0.120
70	1.867	0.117	2.469	0.155	0.140
80	2.134	0.134	2.822	0.177	0.160
90	2.400	0.150	3.174	0.199	0.180
100	2.67	0.17	3.53	0.22	0.20
150	4.00	0.25	5.29	0.33	0.30
200	5.33	0.33	7.05	0.44	0.40
250	6.67	0.42	8.82	0.55	0.50
300	8.00	0.50	10.58	0.66	0.60
350	9.33	0.58	12.34	0.77	0.70
400	10.67	0.67	14.11	0.88	0.80
450	12.00	0.75	15.87	0.99	0.90
500	13.34	0.84	17.64	1.11	1.00
600	16.00	1.00	21.16	1.33	1.20
700	18.67	1.17	24.69	1.55	1.40
800	21.34	1.34	28.22	1.77	1.60
900	24.00	1.50	31.74	1.99	1.80
1000	26.67	1.67	35.27	2.21	2.00

司馬秤換公制

司馬兩	司馬斤	克
1		37.5
2		75
3		112.5
4	0.25	150
5		187.5
6		225
7		262.5
8	0.5	300
9		337.5
10		375
11		412.5
12	0.75	450
13		487.5
14		525
15		562.5
16	1	600
24	1.5	900
32	2	1200
40	2.5	1500
48	3	1800
56	3.5	2100
64	4	2400
80	5	3000

英制換公制

安士	磅	克
1		28.5
2		57
3		85
4	0.25	113.5
5		142
6		170
7		199
8	0.5	227
9		255
10		284
11		312
12	0.75	340.5
13		369
14		397
15		426
16	1	454
24	1.5	681
32	2	908
40	2.5	1135
48	3	1362
56	3.5	1589
64	4	1816
80	5	2270

容量

量杯	公制（毫升）	美制（液體安士）
1/4 杯	60 ml	2 fl. oz.
1/2 杯	125 ml	4 fl. oz.
1 杯	250 ml	8 fl. oz.
1 1/2 杯	375 ml	12 fl. oz.
2 杯	500 ml	16 fl. oz.
4 杯	1000 ml /1 公升	32 fl. oz.

量匙	公制（毫升）
1/8 茶匙	0.5 ml
1/4 茶匙	1 ml
1/2 茶匙	2 ml
3/4 茶匙	4 ml
1 茶匙	5 ml
1 湯匙	15 ml

特色上海食材贊助

三陽泰

專營上海南雜貨食材、金華火腿、大閘蟹、嘉湖糉、江南特色蔬菜及醬料

訂貨電話：96523632　23636238

鳴謝

Inhesion Asia
Prestige Hospitality Tabletop Specialist

上海菜世家　黃可尚先生

杭幫菜大師　傅月良先生

山東旅游職業學院　趙建文教授

海味乾貨專家　林長治先生

梁渭瑩女士

朱文俊先生

王彩燕女士

作者簡介

　　陳紀臨、方曉嵐夫婦，是香港著名食譜書作家、食評家、烹飪導師、報章飲食專欄作家。他們是近代著名飲食文化作家陳夢因（特級校對）的兒媳，傳承陳家兩代的烹飪知識，對飲食文化作不懈的探討研究，作品內容豐富實用，文筆流麗，深受讀者歡迎，至今已在香港出版了十多本食譜書，作品遠銷海外及國內市場，更在台灣多次出版。

　　2016年陳紀臨、方曉嵐夫婦應出版商 Phaidon Press 的邀請，用英文為國際食譜系列撰寫了 *China The Cookbook*，介紹全國 33 個省市自治區的飲食文化和超過 650 個各省地道菜式的食譜，這本書得到國際上好評，並為世界各大主要圖書館收藏。這本書的法文、德文、西班牙文已經出版，將會陸續出版中文、意大利文、荷蘭文等，為中國菜在國際舞台上作出有影響力的貢獻。

如有查詢，請登入：

陳家廚坊讀者會

或電郵至：
chanskitchen@yahoo.com

著者	Author
方曉嵐、陳紀臨	Diora Fong, Keilum Chan

責任編輯	Editor
譚麗琴	Catherine Tam

攝影	Photographer
梁細權、幸浩生、葉冲	Leung Sai Kuen, Johnny Han, George Ip

裝幀設計	Design
羅美齡	Amelia Loh

排版	Typography
何秋雲	Sonia Ho

出版者　Publisher
萬里機構出版有限公司
香港北角英皇道499號
北角工業大廈20樓
電話　Wan Li Book Company Limited
20/F, North Point Industrial Building,
499 King's Road, North Point, Hong Kong
Tel: 2564 7511
Fax: 2565 5539
Email: info@wanlibk.com
Web Site: http://www.wanlibk.com
　　　　　http://www.facebook.com/wanlibk

發行者　Distributor
香港聯合書刊物流有限公司
香港荃灣德士古道220-248號
荃灣工業中心16樓
SUP Publishing Logistics (Hong Kong) Limited
16/F, Tsuen Wan Industrial Centre, 220-248 Texaco Road,
Tsuen Wan, NT., HK
電話　Tel: 2150 2100
傳真　Fax: 2407 3062
電郵　Email: info@suplogistics.com.hk

承印者　Printer
美雅印刷製本有限公司
Elegance Printing & Book Binding Co. Ltd.

出版日期　Publishing Date
二零二零年十二月第一次印刷
First print in December 2020

規格　Size
16開（253 mm × 190 mm）
16K (253 mm × 190 mm)